Delineating Organs at Risk in Radiation Therapy

Giampiero Ausili Cefaro
Domenico Genovesi • Carlos A. Perez

Delineating Organs at Risk in Radiation Therapy

Foreword by Vincenzo Valentini

Giampiero Ausili Cefaro
Department of Radiation Oncology
"G. d'Annunzio" University
Chieti, Italy

Domenico Genovesi
Department of Radiation Oncology
"G. d'Annunzio" University
Chieti, Italy

Carlos A. Perez
Department of Radiation Oncology
Mallickrodt Institute of Radiology
Washington University,
St. Louis, USA

ISBN 978-88-470-5852-1
ISBN 978-88-470-5257-4 (eBook)
DOI 10.1007/978-88-470-5257-4
Springer Milan Dordrecht Heidelberg London New York

Softcover reprint of the hardcover 1st edition 2013

Cover design: Ikona S.r.l., Milan, Italy
Typesetting: Grafiche Porpora S.r.l., Segrate (MI), Italy

Springer-Verlag Italia S.r.l. – Via Decembrio 28 – I-20137 Milan
Springer is a part of Springer Science+Business Media (www.springer.com)

Foreword

Optimal decision-making in the diagnosis, treatment and support of cancer patients is increasingly dependent upon multidisciplinary and multidimensional knowledge. As the pathway of care becomes more complex, the potential for miscommunication, poor coordination between providers, and fragmentation of services and knowledge increases. This constitutes a challenge for all medical and health care professionals involved in the management of a specific tumour disease, whose approach to cancer care is guided by their willingness to agree on evidence-based clinical decisions and to co-ordinate the delivery of care at all stages of the process.

Radiation oncology is a flexible, well-understood, organ-sparing and cost-effective component of cancer therapy that has travelled far over the last century. This progress is illustrated by the numerous innovative developments in treatment equipment and the new delivery technologies and associated imaging modalities that collectively have enabled patient access to highly optimised precision radiation therapy; furthermore, there have been remarkable advances in our understanding of the biological basis of radiation effects and, most recently, the emerging use of novel molecularly targeted therapeutics that hold the promise of delivering further substantial improvements in tumour control and patient cure. To allow our discipline to support superb patient care, a continuous effort in both basic education and the acquisition of new knowledge is mandatory.

One of the more demanding components of modern radiation oncology practice is the definition of irradiation volumes. Inconsistencies in contouring the target and critical structures can seriously undermine the precision of conformal radiation therapy planning and are generally considered to be the biggest and most unpredictable source of uncertainties in radiation oncology. Imaging technology in modern radiation therapy is constantly supporting the integration of new and improved imaging modalities in the contouring process.

Furthermore, we are approaching an era with more organ-preserving treatments and more elderly patients. The prevention of acute and late effects will become a key consideration in the treatment choice.

A group of closely knit colleagues dedicated to the practice of radiation oncology at the University of Chieti in Italy, together with Professor Carlos A. Perez from the Department of Radiation Oncology at the Mallinckrodt Institute of Washington University in St. Louis (USA), have produced this comprehensive book entitled *Delineating Organs at Risk in Radiation Ther-*

apy. They have taken advantage of their extensive and constantly developing clinical experience and their scientific knowledge in order to identify possible solutions to the growing challenges in the field of radiation oncology and offer effective guidance in the delineation of organs at risk.

The book is divided into three parts and is the product of unique contributions from twelve chapter co-authors. The first part offers a thorough synopsis of the anatomy and physiopathology of radiation-induced damage. The second part, with valuable contributions from US co-authors, addresses how to take into consideration the biology and the modelling of normal organs as well as the best imaging practice for volumetric acquisition of all anatomical regions. The third part offers an effective gallery of individual organs at risk in each anatomical region on axial CT scans.

It is easy to predict that many radiation oncologists will find this book to be a realistic, dynamic and well-documented source of information. It will contribute in introducing robust semantics into the language of our community, in correctly managing the irradiation of different tumour sites and in introducing effective and sharable quality assurance programmes into our practice, including in the delineation process. Above all, many of our patients will benefit from the improvements in the practice of radiation oncology that this textbook will foster.

Having had the privilege of long acquaintance with the editors of this book, it is easy for me to identify their deep sense of dedication to their professional life, and I am sure that this will also be readily apparent to all readers. It is well summarised in the following sentence: "Deeds without knowledge are blind, and knowledge without love is sterile" (Benedetto XVI, *Caritas in Veritate*, chapter 30, 2010).

I am pleased to express my best compliments and gratitude to the editors and to all the other authors for the accomplishment of this work based on a robust multidisciplinary and multidimensional knowledge platform.

May 2013

Vincenzo Valentini
Professor and Chair
Department of Radiotherapy
Università Cattolica del Sacro Cuore
Rome, Italy

Contents

Acknowledgements

We would like to gratefully acknowledge the valuable contribution given to this book by the following collaborators:

Antonietta Augurio[°]
Raffaella Basilico[*]
Massimo Caulo[*]
Angelo Di Pilla[°]
Monica Di Tommaso[°]
Bahman Emami[§]
Antonella Filippone[*]
Rossella Patea[*]
Maria Taraborrelli[°]
Marianna Trignani[°]
Lucia Anna Ursini[°]
Annamaria Vinciguerra[°]

[°] Department of Radiation Oncology, University Hospital, Chieti, Italy
[*] Department of Radiology, University Hospital, Chieti, Italy
[§] Department of Radiation Oncology, Loyola University Medical Center, Chicago, USA

Chapter 1
Introduction

Modern 3D conformal radiotherapy (3D-CRT) and its technological developments (intensity-modulated radiation therapy, image-guided radiation therapy, stereotactic radiation therapy) permit administering the prescribed radiation dose to the target volume, applying dose-escalation regimens, while limiting dose to healthy tissues.

The advances of radiation techniques increase the need for both greater accuracy in delineating gross tumor volume (GTV), clinical target volume (CTV), and organs at risk (OARs), and more accurate quantitative evaluation of the dose delivered to the CTV and OARs. For these reasons, delineating GTV, CTV, and – especially – OARs plays a fundamental role in the radiation treatment planning.

From planning to delivery of radiation treatment, radiological imaging is indispensable for identifying clinical volumes. The phase of planning is traditionally based on computed tomography, often performed without contrast, but this method is rather limited for defining some OARs, especially structures with similar electron density to adjacent structures.

Associated morphological (computed tomography, CT, magnetic resonance, MR) and functional imaging (positron emission tomography, PET, single photon emission computed tomography, SPECT, functional magnetic resonance, fMR) can optimize volumes delineation and the collaboration between the radiation oncologist, the radiologist and the nuclear physician, in order to interpret accurately individual imaging methods, is necessary.

These rationales have been the starting point of this book. This textbook is divided into three sections. The first section briefly recalls anatomy and physiopathology of radiation-induced damage. The second section describes modeling, radiation dose constraints for organs at risk according to the most recent evidence and technical notes for volumetric acquisition of OARs in different anatomical districts. The third section identifies individual OARs of four anatomical regions (brain, and head and neck, mediastinum, abdomen, pelvis) on axial CT scans; for each, OAR anatomo-radiological limits (craniocaudal, anteroposterior, and laterolateral) are shown and reported in special summary tables and on CT scan images, together with an iconography of OARs.

This work represents a multidisciplinary model aimed at improving quality assurance in the delicate phase of contouring. However, it is desirable that each radiation therapy center adopt its own model, comparing it with experiences reported in the literature, with the perspective of profitable growth and conscious interactivity.

G. Ausili Cefaro, D. Genovesi, C.A. Perez, *Delineating Organs at Risk in Radiation Therapy*,
DOI: 10.1007/978-88-470-5257-4_1, © Springer-Verlag Italia 2013

Part I

Anatomy and Physiopathology of Radiation-Induced Damage to Organs at Risk

Chapter 2.1
Brain, Head and Neck

2.1.1 Brain and Brainstem

The brain is contained within the skull and is enveloped by membranes called meninx, dura mater, arachnoid membrane, and pia mater. Extensions of the dura mater stabilize the position of the brain, forming structures such as the falx cerebri, tentorium cerebelli, cerebellar falx, and sella diaphragm. Cerebrospinal fluid (CSF), or liquor, produced in the choroid plexuses, protects the delicate nerve structures, sustains the brain, and conveys nutritive and chemical substances and waste products. It flows continuously from ventricles of the central canal of the spinal cord into the subarachnoid space. Diffusion through the arachnoid granulations in the upper sagittal sinus conveys the CSF to the venous bloodstream. In this complex system, the hematoencephalic barrier keeps neural tissue isolated from the general bloodstream. Neural tissue is untouched in the entire central nervous system (CNS), with the only exception being small parts of the hypothalamus, pineal body, and choroid plexus in the membranous roof of the diencephalon and the medulla oblongata.

Consciousness, memory, rationality, and motor functions are the main tasks controlled by the brain, which consists of six regions: telencephalon, diencephalon, mesencephalon, cerebellum, pons, and medulla oblongata (spinal bulb) (Fig. 2.1.1); the medulla oblongata, pons, and mesencephalon form the encephalic trunk [1]:

- The telencephalon and diencephalon, entirely contained within the skull, constitute the cerebrum. The surface of the telencephalon is irregular because of convolutions, or gyri, which are separated by grooves, or fissures. The longitudinal fissure separates the two cerebral hemispheres. In addition, fissures divide the telencephalic cortex into lobes: frontal, parietal, temporal and occipital. The diencephalon links the cerebral hemispheres and encephalic trunk. It comprises the epithalamus, the two thalami, and the hypothalamus. In the posterior region, it contains the pineal gland, or epiphysis.
- The mesencephalon, the shortest portion of encephalic trunk, is about 2–cm long and lies within the posterior cranial fossa.
- The cerebellum, situated in the posterior cranial fossa, has a 15-cm-long transverse axis and is 3-cm thick in the area of the vermis and 5-cm thick in the area of the two cerebellar hemispheres. The cerebellum is characterized by an irregular surface and is divided into two lobes (anterior and posterior), vermis cerebelli, and a flocculonodular lobe.
- The pons is situated between the mesencephalon and medulla oblongata. It is 27-mm long and 38-mm wide and forms a prominent protrusion on the front surface of the encephalic trunk. It also constitutes the floor of the fourth ventricle and is connected to the cerebral hemispheres, which lie posterior to it.
- The medulla oblongata links the brain to the

G. Ausili Cefaro, D. Genovesi, C.A. Perez, *Delineating Organs at Risk in Radiation Therapy*,
DOI: 10.1007/978-88-470-5257-4_2-1, © Springer-Verlag Italia 2013

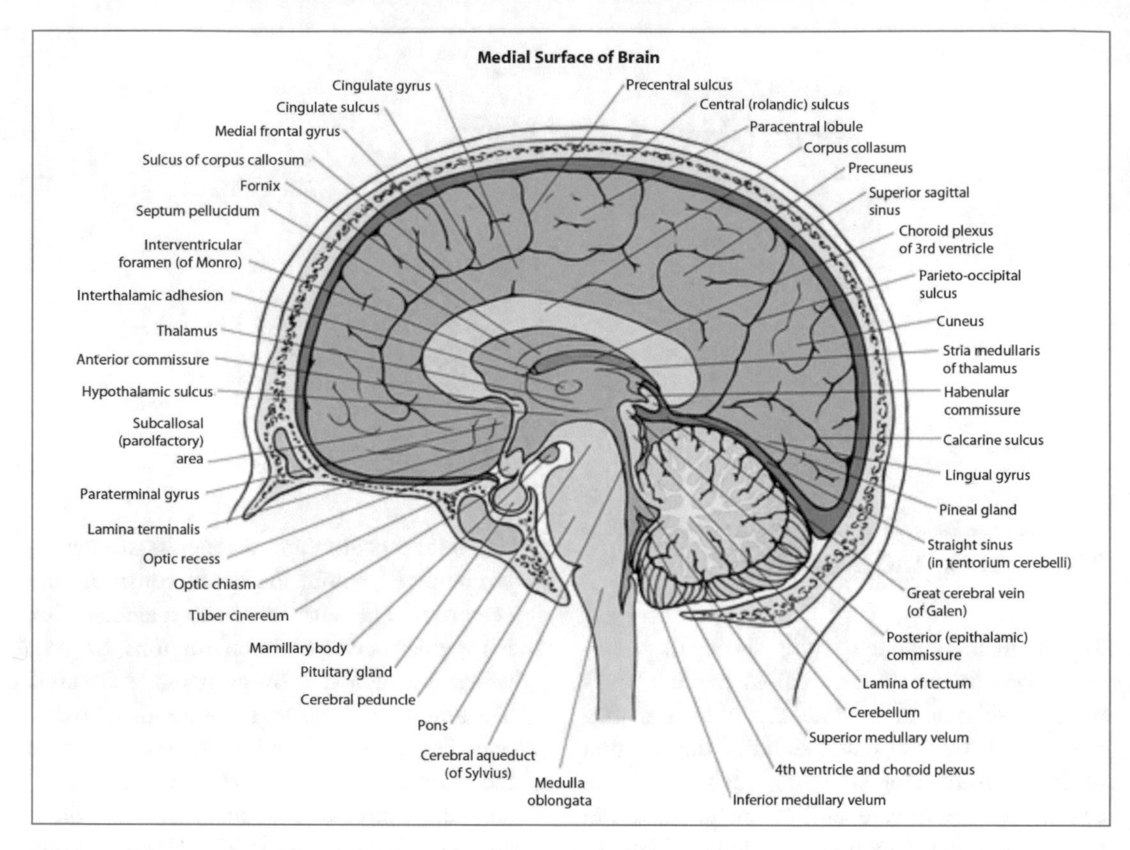

Fig. 2.1.1 Brain anatomy. This drawing of the medial surface of the brain shows the relationship of the brain to the skull. Note that the brain virtually occupies the entire volume of the skull, except for small portions of space left for the vascular system and cerebrospinal fluid. Reproduced with permission from [60]

spinal cord. It is a 3-cm-long and 2-cm-wide stretched structure situated in the posterior cranial fossa, with its lower surface following the slope of the occipital bone and at its posterior connected to the cerebellum by the inferior cerebellar peduncles.

Radiation-induced histological alterations in the brain can be clinically summed up as follows:

- loss of parenchymatous cells, which implies the demyelination of white matter, encephalomalacia, and neuronal loss;
- vascular endothelial injury, which in the acute stage may cause altered permeability, with interruption of the hematoencephalic barrier; in a later phase, telangiectasis, hyalinosis, and fibrinoid deposit in vessels occur.

In cerebral tissue, white matter necrosis is the most frequent histopathological consequence of exposure to a high level of radiation; the endothelium or glial stem cells are the main targets of damage [2]. Vascular alteration is principally responsible for such damage, with significant white matter demyelination and necrosis. In addition to these alterations in cerebral tissue architecture, mutations in cell composition, with an increase in reactive astrocytes (gliosis), can be observed; this produces free oxygen radicals, cytokines, and growth factors, causing inflammatory damage [3, 4]. It appears that the excessive production of free radicals is involved in the development of late damage [2–4].

In all, radiation-induced damage to cerebral tissue is the consequence of complex interactions, including loss of endothelial cells and increased production of free radicals and inflammatory mediators (Table 2.1.1).

Table 2.1.1 Physiopathology of cerebral damage

Anatomical structure	Damage mechanism
Cerebral parenchyma	Increasing number of astrocytes (reactive gliosis) → production of free radicals and inflammatory mediators → inflammatory harm
Endothelium	Alteration of vessel permeability and hematoencephalic barrier → demyelination Loss of endothelial cells as a consequence of the increase in free radicals (acute phase) Hyalinosis and fibrinoid deposit (late phase)

2.1.2 Eye

The apparatus of sight comprises: the eye, or eyeball, situated in the eye socket; the visual pathway, the neural tract that extends from the eyeball to the cerebral cortex; and the adnexa oculi, which are situated all around the eyeball and are divided into a motor apparatus and a protection apparatus. The eyeball consists of two spherical segments with different bending radii: the anterior segment is the cornea, and the posterior segment is the sclera. The eye can be considered an appendix of the brain. The surface of the eyeball consists of three layers:

- external tunica: fibrous tunic around the eye bulb is represented anteriorly by the cornea, which is transparent, and posteriorly by the sclera, which is opaque;
- medium tunica or uvea: vascular, pigmented layer that anterior to posterior comprises the iris, the ciliary body, and the choroid;
- internal tunica: neuronal layer represented by the retina.

These three layers are interrupted posteriorly to allow retinal nerve fibers to create a bundle and form the optic nerve.

There are three cavities within the eyeball. The anterior chamber, between the cornea and iris, contains a clear liquid (aqueous humor). The posterior chamber, behind the iris, also contains a clear liquid (aqueous humor). The vitreous chamber, the largest, contains a gelatinous liquid (vitreous humor). Between the anterior and vitreous chambers is the crystalline lens, which is solid and transparent. It is part of the dioptric system and adjusts its refractory capacity according to the changing distance of an object by modifying its bending radius (Fig. 2.1.2) [5]. When light

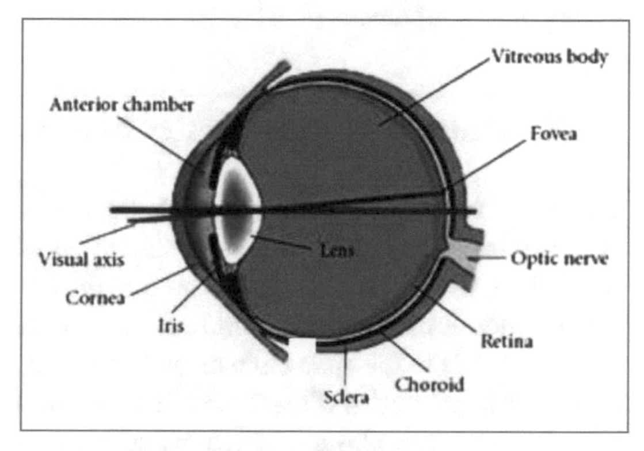

Fig. 2.1.2 Anatomical picture of the human eye. Reproduced with permission from [56]

travels through the cornea, the anterior chamber, pupil, crystalline lens, and vitreous humor, it is directed toward the retina where it crosses the entire retinal layer and hits photoreceptors, which are in direct contact with neuronal cells, the function of which is to transmit the visual impulse [6].

Radiation-induced damage to the eye varies according to the structure of the eye bulb involved. Skin, mucosa, and glands of the visual apparatus respond to radiation in the same way as skin and adnexa of other parts of the body. Acute events may include erythema of the eyelid, conjunctivitis, reduced lacrimal gland secretion, and other damage. At radiation doses <50 Gy, these effects are usually temporary and reversible, whereas higher doses may cause more severe alterations, which may even be irreversible [7].

As the crystalline lens is composed of cells that undergo a regular division cycle, it is highly sensitive to radiation. It does not show signs of acute toxicity, but cataracts occur within 2 or 3 years. Cataracts represent damage to the germinal epithelium of the lens, with consequent cell

Table 2.1.2 Physiopathology of ocular damage

Anatomical structure	Damage mechanism
Germinal epithelium of the lens	Apoptosis and mitosis → cataract

death and compensation mitosis (Table 2.1.2). With regard to the retina, choroid, and optic nerve papilla, most damage is chronic and occurs at the vascular level, causing ischemia of the posterior intraocular components [7, 8].

2.1.3 Optic Nerve and Optic Chiasm

The optic nerve, the second encephalic nerve, is an exclusively sensory nerve, the function of which is to convey visual stimuli through the optic pathway. It originates in ganglion cells in the retina, extends to the optic chiasm, and continues to the occipital cortex (Fig. 2.1.3). It is covered by three connective layers, which are continuous with the encephalic meninges. The optic nerve consists of four segments:

- intrabulb segment: very short; corresponds to the part of the optic nerve that crosses the choroid and sclera cribriform layer;
- intraorbital segment: longest section; travels through the eye socket from the posterior region of the eyeball up to the optic canal of the sphenoid bone;
- intraocular segment: short; positioned in the optic canal of the sphenoid bone;
- intracranial segment: short; stretches from the optic canal of the sphenoid bone to the optic chiasm.

The optic chiasm is a white, rectangular layer with a transverse major axis, top down, inclined in an anteroposterior direction. The optic nerves, originating in the eyeball, travel to the two front angles; from the two back angles, the optic tract branches off. It rests on the sellar tubercle and on

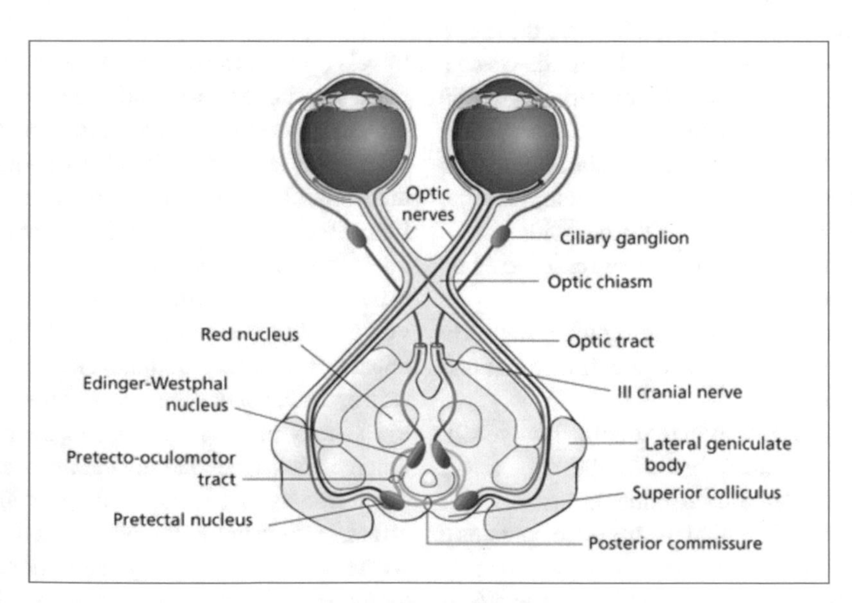

Fig. 2.1.3 Pupillary light reflex. The pupillary afferent axons are identical or lie close to the visual impulse fibers of the ganglion cells. Afferent fibers travel through the optic nerves, chiasm, and tracts, and leave the posterior portion of the optic tract and enter the midbrain, where they synapse with the pretectal nucleus. The axons from the pretectal nucleus then project bilaterally to the Edinger-Westphal nuclei. Note this hemidecussation of the fibers of the pretectal nucleus with crossed fibers at the posterior commissure. The crossed fibers of these intercalated neurons of the pretectal nucleus slightly outnumber the uncrossed fibers. The parasympathetic efferents controlling the pupillary light reflex originate from the Edinger-Westphal cell group of the oculomotor nucleus and synapse within the ciliary ganglion (which lies in the orbit within the muscle cone). Reproduced with permission from [61]

Table 2.1.3 Physiopathology of optic chiasm and optic nerve damage

Anatomical structure	Damage mechanism
Neuron	Production of free radicals and inflammatory mediators → DNA damage and inflammatory alterations → loss of axons
Endothelium	Loss of endothelial cells as a consequence of an increase in free radicals (acute phase) Hyalinosis and fibrinoid deposits (late phase)

the front side of the sellar diaphragm. Posteriorly it makes contact with the lamina terminalis and tuber cinereum; on both sides, it makes contact with the perforated substance [9].

The mechanism of radiation-induced optic neuropathy, with its consequent loss of sight, remains unclear. It might be due to lesions of the retrobulbar optic nerve, the optic chiasm, or the retrogeniculate pathway. It occurs about 18 months after radiation treatment and after a cumulative dose >50 Gy or a single dose to the visual apparatus >10 Gy. Radiation-induced damage at neuro-ophthalmic level appears mainly like late toxicity. Its predominant mechanism of action is alteration to the white matter. This process starts with the production of free radicals as a consequence of DNA alterations to normal tissue cells, which is caused by ionizing radiation. It seems that the primary site of cell damage is the vascular endothelium and parent cells in neuroglia. Even in the terminal stage of radiation-induced optic neuropathy, the main target is represented by the endothelium and glial cells; at this stage the damage is characterized by reduced blood vessel caliber and occlusion, axonal loss, demyelination, and presence of exudates with fibrin [10–12]. Whatever the exact cell pathogenesis, tissues reveal hypovascularization, hypocellularity, and hypoxia (Table 2.1.3).

2.1.4 Ear

The ear, the organ responsible for the hearing function, is divided into three parts: outer, middle, and inner. The outer ear consists of the pinna and ear canal, or auditory meatus. The pathway of the ear canal is not perfectly straight and travels for about 24 mm to the tympanic membrane, or eardrum. The canal is a fibrous–cartilaginous structure for one third of its lateral side; the remaining two thirds is a bony structure. Lengthwise, it is covered by the skin. The middle ear contains the eardrum, behind which lies the tympanic cavity – an air space situated within the temporal bone. It also contains three small bones stapes (stirrup), incus (anvil), and malleus (hammer), which form the chain of auditory ossicles. The middle ear connects with the pharynx (nasopharynx) via the Eustachian tube and with the mastoid cavities posteriorly. The inner ear is the deepest section and is encased in the temporal bone. It consists of the auricular labyrinth (or bony labyrinth) and is divided into two parts: the organ of hearing (cochlea) and the organ of balance (vestibular apparatus, which consists of three semicircular canals and the vestibule) (Fig. 2.1.4).

In the physiology of hearing, each of the three parts of the ear performs a specific function. The outer ear collects, conveys, and amplifies the sound waves, which cause the auricular labyrinth to vibrate toward the middle ear. Inside the middle ear, the energy of the sound waves is transformed into mechanic vibrations of the bony structure of the middle ear (movement of the three ossicles). This movement, both at the front and at the back within the space of the oval window of the cochlea, causes the stapes to transmit the kinetic impulse to the perilymph, which is extracellular fluid contained within the cochlea. Through the endolymph (fluid contained in the membranous labyrinth) in the cochlear duct, sound waves are transmitted from the scala vestibuli to the scala tympani, causing vibration of the membranes that separate the scalae, of cochlea. This stimulates hair cells of the auditory system,

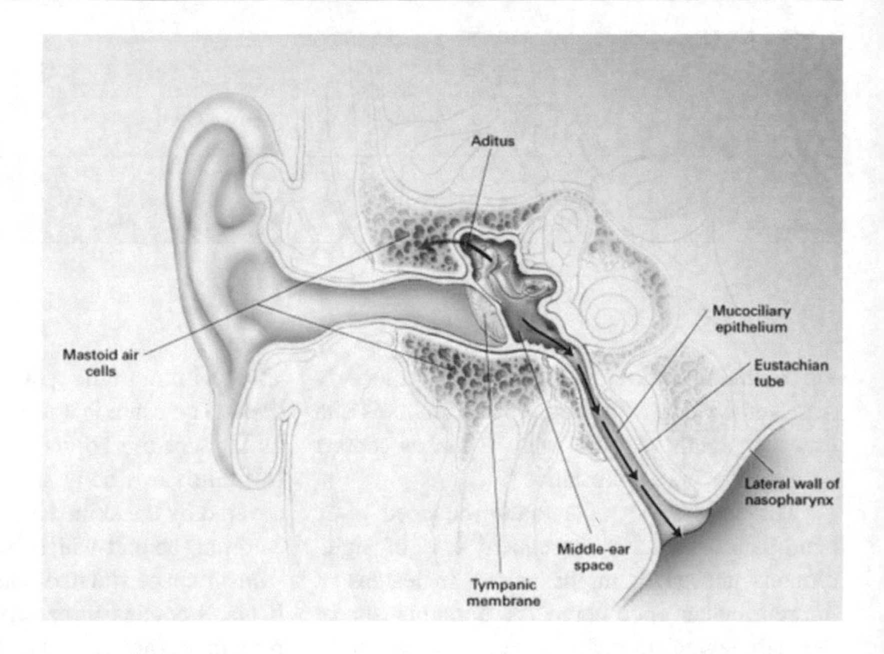

Fig. 2.1.4 Anatomical configuration of the ear. Reproduced with permission from [57]

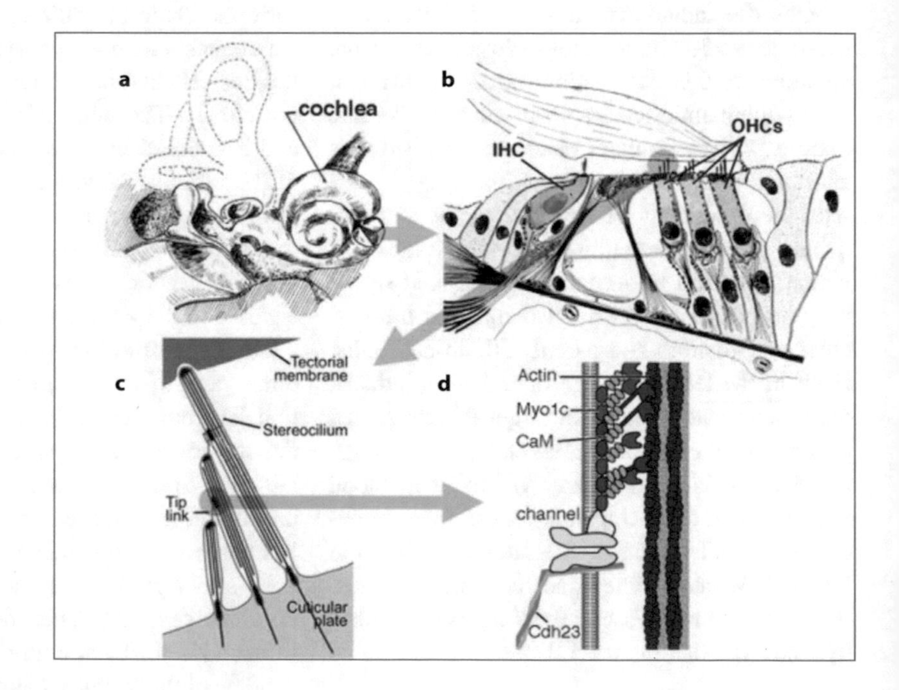

Fig. 2.1.5 Anatomical details of inner ear, cochlea and organ of Corti, the sense organ of mammalian hearing. Reproduced with permission from [58]. *CaM* cell adhesion molecule, *Cdh23* cadherin-23, *IHC* inner hair cell, *OHC* outer hair cell

producing an electrical signal that is transmitted along the acoustic nerve to the auditory cortex (Fig. 2.1.5).

Exposure of the hearing mechanism is inevitable during radiation therapy for cerebral or head and neck tumors. Among the several toxicities induced by radiation therapy, neurological complications and hearing loss are the most serious. Despite the relatively large amount of scientific research carried out on both animals and humans, data about incidence, type, and severity of the induced toxicity are scarce [13].

Radiation-induced damage may affect every structure of the ear (outer, middle, inner, and up

Table 2.1.4 Physiopathology of ear damage

Anatomical structure	Damage mechanism
Endothelium	Atrophy of sensory neuronal cells of inner ear Fibrosis and ossification of fluid spaces
Nerve structures	Vascular damage → atrophy and degeneration of nerve structures Production of inflammatory mediators → edema → compression of cochlear nerve in the ear canal

to the auditory canal). In the outer ear, radiation toxicity causes reactions that involve the skin and cartilage of the preauricular region, the pinna, and the outer auditory canal. Acute or late events may occur, with various rates of morbidity. In the middle ear, the most common complication is Eustachian tube dysfunction as a consequence of otitis media, and temporary loss of hearing. Thickening of the tympanic membrane with sclerosis and perforation has been reported. A high radiation dose may cause fibrosis in the middle ear and atrophy of ossicles. In the inner ear, radiation toxicity may give rise to a wide range of symptoms, such as tinnitus, labyrinthitis, vertigo with balance alteration, and sensorineural deafness.

Radiation-induced hearing loss can be divided into conductive or sensorineural, depending on whether the cause is damage to parts of the middle ear, the cochlea, or the retrocochlear. Vascular insufficiency (alteration of endothelium) has been proposed as the etiology of sensorineural hearing loss. Injury to vascular structures of the inner ear may cause progressive degeneration and atrophy of sensory structures and fibrosis and ossification of fluid spaces in a matter of a few weeks or months after irradiation. Neuronal components may also be damaged, which manifests as degeneration and atrophy of Corti's ganglion and the cochlear nerve. Furthermore, both inflammation and edema may compress the cochlear nerve in the ear canal, causing damage [14, 15] (Table 2.1.4).

2.1.5 Salivary Glands

The salivary glands are divided into major and minor. Major salivary glands are represented by three large glands: parotid, submaxillary or submandibular, and sublingual.

The parotid glands are the largest. They are situated in the lateral region of the neck, called the parotid region, or parotid space, under the pinna and the external auditory meatus posteriorly to of the mandibular ramus and at the front of the sternocleidomastoid muscle. They are connected to the oral cavity through Stensen's duct, which terminates in the cheek close to the superior molar. Each parotid gland is covered by the parotid fascia, in which a superficial side, just against the gland and the skin, and a deeper part, which covers the outer surface of the parotid space, can be distinguished. The superficial and deep fusion of the parotid fascia membranes form the interglandular septum, which separates the parotid into superficial and deep segments. One of the most important anatomical connections of the parotid gland is the one with the facial nerve, which, emerging from the skull through the stylomastoid foramen, crosses the gland, dividing it into its two main branches: the temporal–facial and the cervical–facial and these, in turn, into terminal branches (Fig. 2.1.6).

The submandibular glands are situated in the suprahyoid region inside the submandibular spaces and are encased in a capsule. Wharton's duct emerges from the middle face of the gland and travels to the sublingual sulcus, where it branches out near the lingual caruncle.

The sublingual glands, situated in the sublingual space, appear as an agglomerate of small lobules that often remain separate, being provided with small excretory ducts (Walther's sublingual canals). One of these lobules, larger than the others, is called the major sublingual gland. The main sublingual duct (Bartholin's duct), the function of which is to drain the main gland, opens into the caruncle on the submandibular duct.

The main function of the salivary glands is to secrete saliva, which plays an important role in

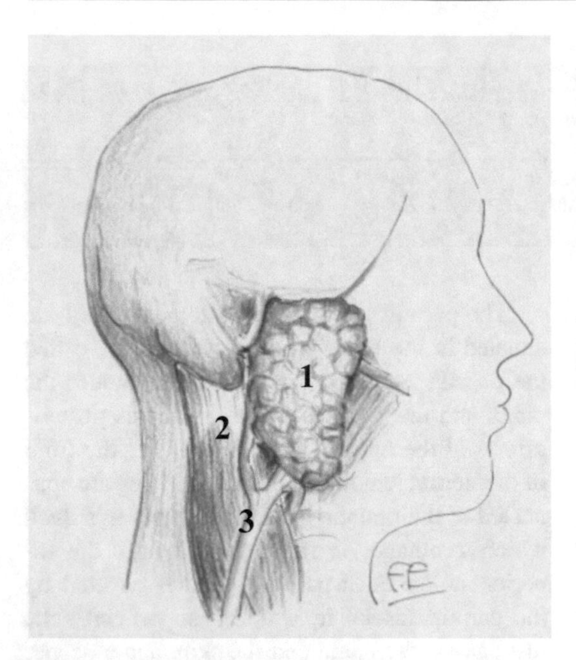

Fig. 2.1.6 Anatomical picture of the parotid glands. *1* Parotid gland; *2* sternocleidomastoid muscle; *3* jugular vein. Drawing by Franca Evangelista

digestion and ensures antibacterial action inside the oral cavity. Saliva is a combination of mucosal and serous secretions composed of water (approximately 94%) and in a variable quantity of ptyalin, mucin, sulfonate, chlor-alkali, sodium phosphate, calcium salts, magnesium, and cellular elements [16].

Functional alteration of the salivary glands is called xerostomia. Radiation-induced xerostomia is a toxicity that appears quite early. A reduction in the salivary output usually begins during the first week of therapy; after 7 weeks of therapy with conventional fractionation, a reduction of 20% in salivary output is registered [17].

There are two hypotheses to explain the mechanism of radiation-induced damage. The first is so-called granulation, in which membranes of secretory granules in acinar cells are damaged by peroxidation of lipids, which is induced by radiation. Consequently, the proteolytic enzymes begin to discharge from the granules causing fast cell lysis. This mechanism seems to show that the volume of the glands remains the same but the secretory function is jeopardized. The second hypothesis considers two different mechanisms: (a) failure in cellular function due to selective damage to the membrane at the expense of the water-secretion receptor; (b) death of stem cells, with consequent inhibition of cell renewal [18].

Apart from the mechanism that determines it, decline in salivary function continues for several months after the end of radiotherapy. A fair recovery may occur from 12 to 18 months after radiotherapy, depending on radiation dose and treated volume of the gland; however, it is only a partial recovery, with a possible 30 % increase in salivary function within 5 years. Therefore, xerostomia is virtually irreversible [19–20] (Table 2.1.5).

2.1.6 Constrictors of the Pharynx

The pharynx is part of both the digestive and respiratory tracts; it is situated behind the cavities of the nose, mouth, and larynx and extends to the esophagus. Furthermore, the Eustachian tubes of the ear originate in the pharynx. The front of the pharynx is somewhat incomplete, with connections with the nasal cavities through the choanae, with the mouth through the isthmus faucium, and with the larynx through the orifice or laryngeal aditus pass (Fig. 2.1.7). The wall of the pharynx, from the outside in, consists of the following strata: tunica adventitia; striated muscle tunica; fibroelastic tunica or pharyngeal fascia or pharyngobasilar fascia; and mucous tunica. Muscles are divided into constrictors and levators.

Table 2.1.5 Physiopathology of gland damage

Anatomical structure	Damage mechanism
Acinar cells	Peroxidation of lipids → harm to membranes of secretory cells → cell lysis Selective damage to the membrane at the expense of the water-secretion receptor
Stem cells	DNA damage → cell death

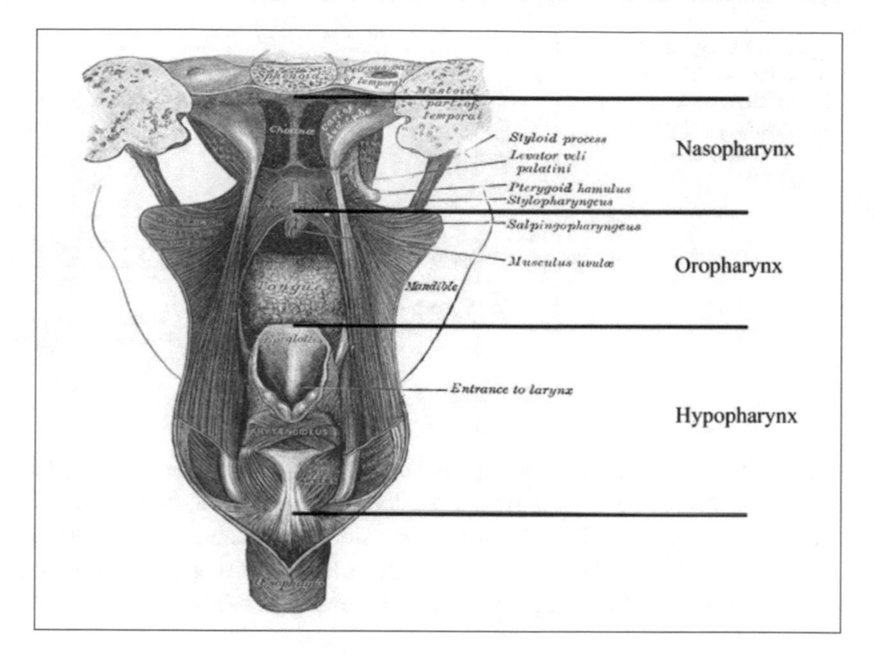

Fig. 2.1.7 Anatomical picture of the anterior of the pharynx. Reproduced with permission from [62]

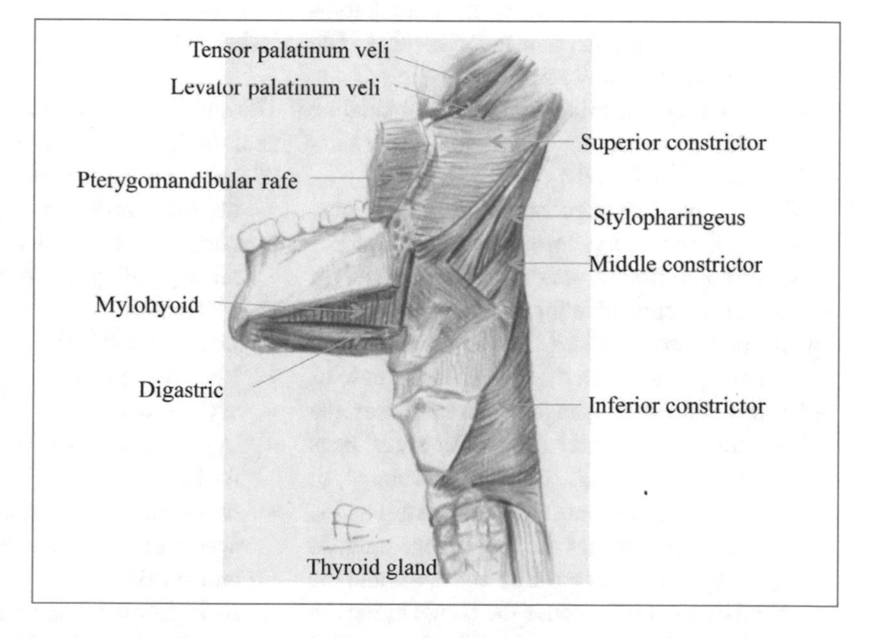

Fig. 2.1.8 Schematic picture of the swallowing muscles. Drawing by Franca Evangelista

Constrictors are represented by superior, medial, and inferior constrictors:

- The bundle of superior constrictors derives from the medial pterygoid plate of the sphenoid bone, the pterygomandibular raphe, the mylohyoid line, the mandible, and from bundles that extend sideways from the radix linguae through the genioglossus. These are quadrilateral bundles and insert into a medial raphe in the posterior side of the pharynx.
- The medial constrictor is a triangular-shaped muscle, the base of which coincides with the medial raphe and its apex with the hyoid bone.
- The inferior constrictor, the largest, is a trapezoid muscle shaped by bundles that arise from the oblique line of the thyroid cartilage and insert posteriorly into the pharyngeal raphe (Fig. 2.1.8).

The constrictor muscles control the swallow-

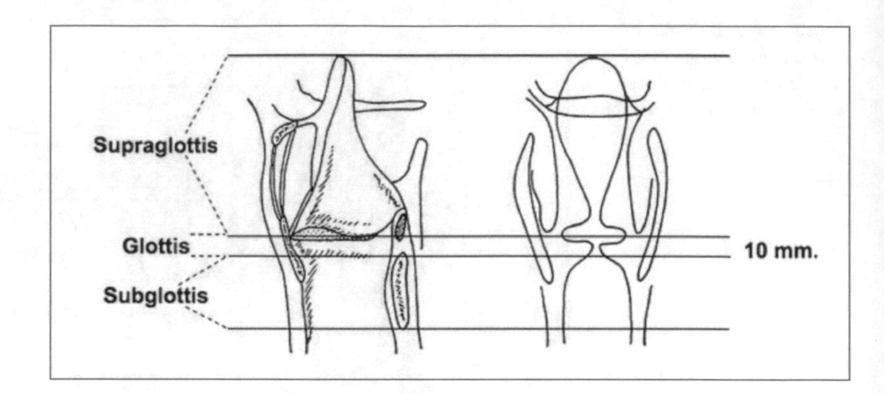

Fig. 2.1.9 Diagram of the larynx showing the different sites and subsites. Reproduced with permission from [59]

ing apparatus: the constrictor superior constricts the epipharynx and levator on the posterior of the pharynx; the medial constrictor constricts the oropharynx; and the inferior constrictor constricts the laryngeal part of the pharynx and levator of the larynx.

Levators are represented by the stylopharyngeus and pharyngopalatinus muscles [21]. The pharyngopalatinus muscle is the levator of the pharynx and larynx, brings the pharyngopalatine arches closer to each other, and dilates the Eustachian tube (Fig. 2.1.9).

Patients who undergo radiation therapy may present inflammation, fibrosis, edema, and necrosis of some tissues, among which are nerves and muscles responsible for oropharyngeal swallowing and movement of the larynx and superior esophageal sphincter [22, 23]. In the case of radiation-induced dysphagia, edema causes the obliteration of structures such as anatomical bags (i.e., vallecula) and ducts (i.e., piriform sinus), so that the food bolus cannot flow downward and is instead conveyed toward the respiratory tract. In late dysphagia, fibrosis prevails on edema and the fibrotic tissue deposits under the skin, among the layers of connective tissue, round muscles, and among muscle fibers [24]. Variables that have a negative influence on radiation-induced dysphagia are smoking during and after radiotherapy, being older, total radiation dose, radiation dose/fraction, interfraction lapse, extended volume of treatment, treatment techniques, weight loss, and position and dimensions of the primary cancer [25–34] (Table 2.1.6).

2.1.7 Larynx

The larynx is an unpaired organ situated in the neck in front of the hypopharynx and consisting of the following structures:

- thyroid cartilage: unpaired, shield-shaped, composed of two laminae in the upper, frontal, and side portions of the larynx;
- epiglottis;
- false vocal folds;
- true vocal folds;
- arytenoids;
- hypoglottis, located under the true vocal folds;
- two commissures, anterior and posterior: respectively the contact points of vocal folds and arytenoids.

In addition to these main cartilages, small, secondary cartilages, e.g., corniculate (Santorini's) , cuneiform (Morgagni's), and other minor cartilages, are mainly situated in the thickness of ligaments (e.g., triticeal cartilages).

Table 2.1.6 Physiopathology of muscle damage

Anatomical structure	Damage mechanism	
Muscle	Inflammatory mediators → edema ↓ Fibroblastic proliferation → fibrosis	Disphagia

Table 2.1.7 Physiopathology of laryngeal damage

Anatomical structure	Mechanism of damage	
Laryngeal mucosa and cartilages	Inflammatory mediators → edema ↓ Fibroblastic proliferation → fibrosis	Dysphonia Dysphagia

The larynx is divided into three levels: supraglottis, glottis, and subglottis:

- The supraglottis is connected to the pharynx through the laryngeal aditus, upward and backward, is roughly oval shaped, and borders the free margin of the epiglottis, the aryepiglottic folds, the arytenoids, and the inter-arytenoid notch. The front consists of the epiglottis, the lateral walls show two projections (false vocal folds) and two cavities (Morgagni's ventricles), and the back corresponds to the laryngeal face of the arytenoids and to the inter-arytenoid notch.
- The glottis lamina consists of a triangular space with a front apex (anterior commissure), virtual in phonation, surrounded by the true vocal folds.
- The subglottis, shaped like an upside-down funnel, enlarges in the skull–caudal direction and is continuous with the trachea.

In the larynx there is a series of muscles that can be distinguished as extrinsic and intrinsic. Intrinsic muscles have both insertions on the larynx; their contraction allows the laryngeal cartilages to move, thus modifying the position of the vocal folds. Intrinsic muscles can be divided into dilator muscles (they allow separation of the vocal folds and opening of the glottis plate), constrictor muscles (which move the vocal folds closer to each other), and tensor muscles of the vocal fold (which increase tension in the vocal folds). Extrinsic muscles insert at the level of the thyroid cartilage, hyoid bone, and sternum, allowing vertical movement of the larynx and contributing to its stabilization, and stabilization of other neck muscles [35].

The larynx has two important functions: breathing and phonatory. Furthermore, it assists with swallowing, thus protecting the respiratory system. The false and true vocal folds contribute to the phonatory function, whereas the epiglottis contributes to the swallowing function. All laryngeal sections play an important role in the respiratory function [35].

As radiotherapy causes inflammation, destroys lymphatic vessels, and later on causes fibrosis, edema of laryngeal structures may appear. In the acute phase, laryngeal edema causes phonation dysfunction, the seriousness of which may vary from moderate aphonia to reduced respiratory space; laryngeal edema may also cause variably serious dysphagia relative to irradiation of the supraglottis and the laryngopharyngeal region. Severity of these dysfunctions is governed by the laryngopharyngeal volume irradiated and dosage. In the later stages, the edema combines with fibrosis, which is responsible for irreversible dysfunction.

Toxicity caused by ionizing radiation may also affect the laryngeal cartilages, but this damage is much more uncommon than laryngeal edema, with which it shares physiopathology, including variable alterations to the chondroradionecrosis of the cartilaginous tissue caused by edema [36, 37] (Table 2.1.7).

2.1.8 Mandible and Temporomandibular Articulation

The mandible is a medial bone in which the lower teeth are located. It is in the shape of "horseshoe", with two posterior, upward-directed branches that form an obtuse angle with the body. On the superior edge of both branches are the upper-concavity notches, which separate two protrusions: anteriorly, the coronoid process, in which the temporal muscle is inserted; posteriorly, the condyle, or condylar process. The latter consists of the head, on which the articular surface is located, and the neck, into which the external pterygoid muscle inserts (Fig. 2.1.10).

The temporomandibular articulation (TMA)

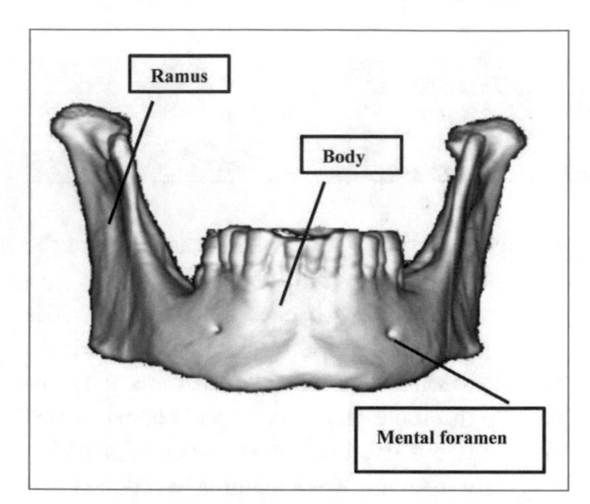

Fig. 2.1.10 Mandible reconstruction, anterior view

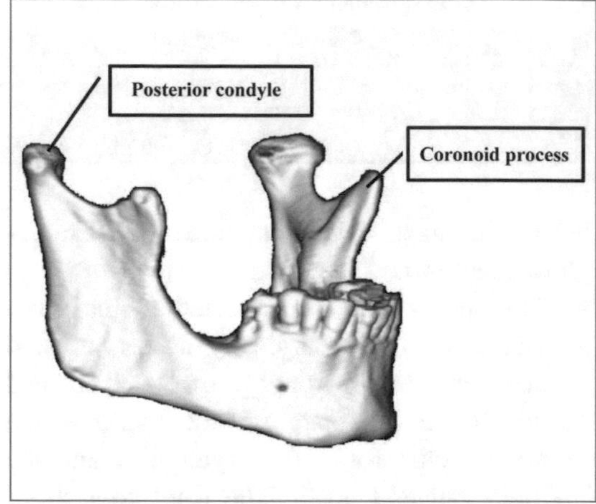

Fig. 2.1.11 Mandible reconstruction, lateral view

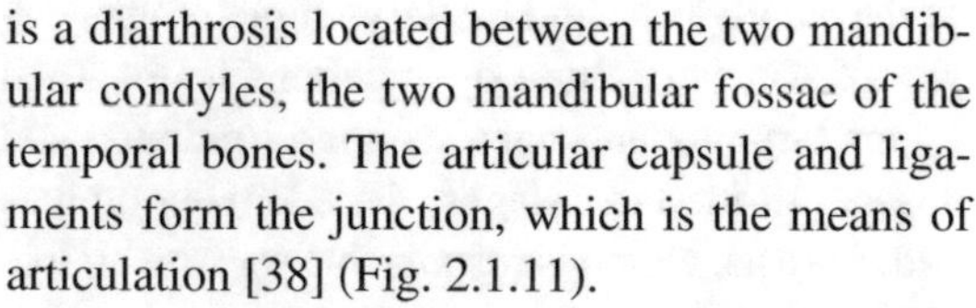

is a diarthrosis located between the two mandibular condyles, the two mandibular fossac of the temporal bones. The articular capsule and ligaments form the junction, which is the means of articulation [38] (Fig. 2.1.11).

As the two TMAs operate simultaneously, they facilitate the lowering and lifting, forward and backward projections, and lateral movements of the mandible, thus permitting phonation, nutrition, and buccal hygiene [38].

In general, the physiopathology of mandibular damage is linked to alteration in nutritional vessels and bony tissue in the same way as in other bony districts. The mandible shows early predisposition to the risk of radiation-induced necrosis due to its vascularization (occlusion of the lower alveolar artery due to the fact that radiation-induced fibrosis is not compensated for by the facial artery) and to a higher bone density in the regions of the molar and premolar teeth, the regions most commonly damaged [39–42].

Furthermore, additional risk elements, such as those attributable to the tumor and to the patient, should be taken into consideration. Elements linked to the tumor are site, stage, and dimensions. Those linked to the patient are tooth extractions following radiotherapy, or surgical operations such as mandibular resection before radiation treatment. All these elements can increase radiation-induced necrosis to the bone and must be kept in mind with regard to damage prevention [42–53].

Trismus, tonic contraction of the masseter muscle, is another serious collateral effect of radiation treatment to the TMA, as it may cause even more serious complications such as malnutrition, phonation dysfunctions, and loss of oral health.

It seems that radiation damage to the TMA is due to a proliferation of fibroblasts with consequent fibroatrophy, which is characterized by cartilage thinning and synovial fluid depletion, and immobilization – even total – of the joint as a result of it. Additionally, pterygoid and masseter muscle injury has been reported. The total dose of radiotherapy seems to be an important determining factor [54, 55].

Acknowledgement This chapter has been written with the contributions of Angelo Di Pilla, Annamaria Vinciguerra, and Marianna Trignani.

References

1. Martini FH, Timmons MJ, Tallitsch RB (2010) Anatomia umana, Edises pp 387–415
2. Rabin BM, Meyer JR, Berlin JW et al (1996) Radiation-induced changes in the central nervous system and head and neck. RadioGraphics 16:1055–1072
3. Kim JH, Brown SL, Jenrow KA et al (2008) Mechanism of radiation-induced toxicity and implications for future clinical trials. J Neuroncol 87:279–286
4. Mayo C, Yorke E, Merchant TE (2010) Radiation-associated brainstem injury. Int J Radiat Oncol Biol Phys 76(3 Suppl):S36–S41

5. Miglior M, Bagolini B, Boles Carenini B et al (1989) Oftalmologia Clinica. Second Edition. Monduzzi, pp 5-13
6. Jeganathan VSE, Wirth A, MacManus MP (2011) Ocular risks from orbital and periorbital radiation therapy: a critical review. Int J Radiat Oncol Biol Phys 79:650–659
7. Marchand V, Dendale R (2010) Normal tissue tolerance to external beam radiation therapy: Eye structures. Cancer Radiothér 14:277–283
8. Gordon KB, Char DH, Sagerman RH (1995) Late effects of radiation on the eye and ocular adnexa. Int J Radiat Oncol Biol Phys 31:1123–1139
9. Cattaneo L (1972) Annotazioni di anatomia dell'uomo. Monduzzi, Vol. 2, pp 278-280
10. Lessell S (2004) Friendly fire: Neurogenic visual loss from radiation therapy. J Neuroophthalmol 24:243–250
11. Danesh-Meyer HV (2008) Radiation-induced optic neuropathy. J Clin Neurosci 15:95–100
12. Mayo C, Martel MK, Marks LB et al (2010) Radiation dose-volume effects of optic nerves and chiasm. Int J Radiat Oncol Biol Phys 76(3 Suppl):S28–S35
13. Jereczek-Fossa BA, Zarowski A, Milani F et al (2003) Radiotherapy-induced ear toxicity. Cancer Treat Rev 29:417–430
14. Bhide SA, Harrington KJ, Nutting CM (2007) Otological toxicity after postoperative radiotherapy for parotid tumours. Clin Oncol (R Coll Radiol) 19:77–82
15. Bhandare N, Jackson A, Eisbruch A et al (2010) Radiation therapy and hearing loss. Int J Radiat Oncol Biol Phys 76(3 Suppl)S50–S57
16. Balboni GC, Bastianini A, Brizzi E et al (1991) Anatomia Umana. Third Edition. Edi Ermes, Vol. 2, pp 64-74
17. Leslie M, Dische S (1991) Parotid gland function following accelerated and conventionally fractionated radiotherapy. Radiother Oncol 22:133-139
18. Chambers MS, Garden AS, Kies MS (2004) Radiation-induced xerostomia in patients with head and neck cancer: Pathogenesis, impact on quality of life and management. Head Neck 26(9):796–807
19. Dirix P, Nuyts S, Van den Bogaert W (2006) Radiation-induced xerostomia in patients with head neck cancer: a literature review. Cancer 107(11):2525–2534
20. Deasy JO, Moiseenko V, Lawrence Marks D (2010) Radiotherapy dose-volume effects on salivary gland function. Int J Radiat Oncol Biol Phys 76(3 Suppl): S58–S63
21. Balboni GC, Bastianini A, Brizzi E et al (1991) Anatomia Umana. Third Edition. Edi Ermes, Vol. 2, pp 83-87
22. Eisbruch A, Lyden T, Bradford CR et al (2002) Objective assessment of swallowing dysfunction and aspiration after radiation concurrent with chemotherapy for head and neck cancer. Int J Radiat Oncol Biol Phys 53:23–28
23. Langmore SE, Krisciunas GP (2010) Dysphagia after radiotherapy for head and neck cancer: etiology, clinical presentation, and efficacy of current treatments. Perspect Swallow Swallow Disord (Dysphagia) 19(2):32–38
24. Russi EG, Corvò R, Merlotti A et al (2012) Swallowing dysfunction in head and neck cancer patients treated by radiotherapy: Review and recommendations of the supportive task group of the Italian Association of Radiation Oncology. Cancer Treat Rev 38(8):1033-49
25. Gramley F, Lorenzen J, Koellensperger E et al (2010) Atrial fibrosis and atrial fibrillation: the role of the TGF-b1 signaling pathway. Int J Cardiol 143(3):405–413
26. Haydont V, Riser BL, Aigueperse J, Vozenin-Brotons M-C (2008) Specific signals involved in the long-term maintenance of radiation-induced fibrogenic differentiation: a role for CCN2 and low concentration of TGF-b1. Am J Physiol Cell Physiol 294(6):C1332–1341
27. Langendijk JA, Doornaert P, Rietveld DHF et al (2009) A predictive model for swallowing dysfunction after curative radiotherapy in head and neck cancer. Radiother Oncol 90(2):189–195
28. Eisbruch A (2004) Dysphagia and aspiration following chemo-irradiation of head and neck cancer: major obstacles to intensification of therapy. Ann Oncol 15:363–364
29. Levandag PC, Teguh DN, Voet P et al (2007) Dysphagia disorders in patients with cancer of the oropharynx are significantly affected by the radiation therapy dose to the superior and middle constrictor muscle: A dose–effect relationship. Radiother Oncol 85:64–73
30. Eisbruch A, Schwartz M, Rasch C et al (2004) Dysphagia and aspiration after chemoradiotherapy for head-and-neck cancer: which anatomic structures are affected and can they be spared by IMRT? Int J Radiat Oncol Biol Phys 60:1425–1439
31. Dornfeld K, Simmons JR, Karnell L et al (2007) Radiation doses to structures within and adjacent to the larynx are correlated with long-term diet and speech-related quality of life. Int J Radiat Oncol Biol Phys 68:750–757
32. Jensen K, Lambertsen K, Grau C (2007) Late swallowing dysfunction and dysphagia after radiotherapy for pharynx cancer: Frequency, intensity, and correlation with dose and volume parameters. Radiother Oncol 85:74–82
33. Fua TF, Corry J, Milner AD et al (2007) Intensity-modulated radiotherapy for nasopharyngeal carcinoma: Clinical correlation of dose to the pharyngoesophageal axis and dysphagia. Int J Radiat Oncol Biol Phys 67:976–981
34. Rancati T, Schwarz M, Allen AM et al (2010) QUANTEC: organ-specific paper. Radiation dose–volume effects in the larynx and pharynx. Int J Radiat Oncol Biol Phys 76(3 Suppl):S64–S69
35. Balboni GC, Bastianini A, Brizzi E et al (1991) Anatomia Umana. Third Edition. Edi Ermes, Vol. 2, pp 292-310
36. Fung K, Yoo J, Leeper HA et al (2001) Vocal func-

tion following radiation for non-laryngeal versus laryngeal tumors of the head neck. Laryngoscope 111:1920–1924
37. RancatiT, Schwartz M, Allen AM et al (2010) Radiation dose-volume effect in larynx and pharynx. Int J Rad Oncol Biol Phys 76(3 Suppl):S64–S69
38. Balboni GC, Bastianini A, Brizzi E et al (1991) Anatomia Umana. Third Edition. Edi Ermes, Vol. 1, pp 93-96
39. Harris M (1992) The conservative management of osteoradionecrosis of the mandible with ultrasound therapy. Br J Oral Maxillofac Surg 30:313–318
40. Marx RE (1983) A new concept in the treatment of osteoradionecrosis. J Oral Maxillofac Surg 41:351–357
41. Ben-David MA, Diamante M, Radawski JD et al (2007) Lack of osteoradionecrosis of the mandible after intensity-modulated radiotherapy for head and neck cancer: likely contributions of both dental care and improved dose distributions. Int J Radiat Oncol Biol Phys 68:396–402
42. Meyer I (1970) Infectious diseases of the jaws. J Oral Surg 28:17–26
43. Marx RE (1983) Osteoradionecrosis: a new concept of its pathophysiology. J Oral Maxillofac Surg 41:283–288
44. Chrcanovic BR, Reher P, Sousa AA, Harris M (2010) Osteoradionecrosis of the jaws. A current overview. Part 1: Physiopathology and risk and predisposing factors. Oral Maxillofac Surg 14:3–16
45. Silvestre-Rangil J, Silvestre FJ (2011) Clinico-therapeutic management of osteoradionecrosis: a literature review and update. Med Oral Patol Oral Cir Bucal 16:e900–e904
46. Oh HK, Chambers MS, Martin JW Et al (2009) Osteoradionecrosis of the mandible: treatment outcomes and factors influencing the progress of osteoradionecrosis. J Oral Maxillofac Surg 67:1378–1386
47. Jereczek-Fossa BA, Orecchia R (2002) Radiotherapy-induced mandibular bone complications. Cancer Treat Rev 28:65–74
48. Bagan JV, Jiménez Y, Hernández S et al (2009) Osteonecrosis of the jaws by intravenous bisphosphonates and osteoradionecrosis: a comparative study. Med Oral Patol Oral Cir Bucal 14:e616–e619
49. Lee IJ, Koom WS, Lee CG Et a l (2009) Risk factors and dose–effect relationship for mandibular osteoradionecrosis in oral and oropharyngeal cancer patients. Int J Radiat Oncol Biol Phys 75(3 Suppl):1084–1091
50. Lyons A, Ghazali N (2008) Osteoradionecrosis of the jaws: current understanding of its pathophysiology and treatment. Br J Oral Maxillofac Surg 46:653–660
51. Marx RE, Johnson RP (1987) Studies in the radiobiology of osteoradionecrosis and their clinical significance. Oral Surg Oral Med Oral Pathol 64:379–390
52. Mendenhall WM (2004) Mandibular osteoradioncrosis. J Clin Oncol 22:4867–4868
53. Wahl MJ (2006) Osteoradionecrosis prevention myths Int J Radiat Oncol Biol Phys 64:661–669
54. Dijkstra PU, Kalk WW, Roodenburg JL (2004) Trismus in head and neck oncology: a systematic review. Oral Oncol 40:879–889
55. Teguh DN, Levendag PC, Voet P et al (2008) Trismus in patients with oropharyngeal cancer: relationship with dose in structures of mastication apparatus. Head Neck 30:622–630

Chapter 2.2
Mediastinum

2.2.1 Humeral Head

The humerus articulates with the scapula superiorly and the radius and ulna inferiorly and consists of a body and two limbs: proximal and distal. The proximal limb is thickened and joins the body at the surgical neck. Its articular surface, the humeral head, is wide, almost hemispherical, and covered with cartilage. On the upper margin of the humeral head is a slight narrowing, the anatomical neck, and two protrusions, the greater and lesser tubercles, or tuberosities. Between the tuberosities is a groove, the bicipital sulcus, through which pass the two crests and tendon of the long head of the bicipital muscle. The pectoralis major muscle inserts on the lateral edge of the humeral head, and the large dorsal muscle and large round muscle insert onto the medial brim [1] (Fig. 2.2.1).

Radiation effects on bones vary as a consequence of dose, ray energy, and fractionation. The pathological effects of radiations differ from the immature (growing) to the mature (adult) skeleton. In an immature skeleton, radiation affects chondrogenesis and resorption of calcified cartilage [2]. In an adult skeleton, the effects of radiation occur mainly on osteoblasts, with consequent reduction in bone formation [3, 4].

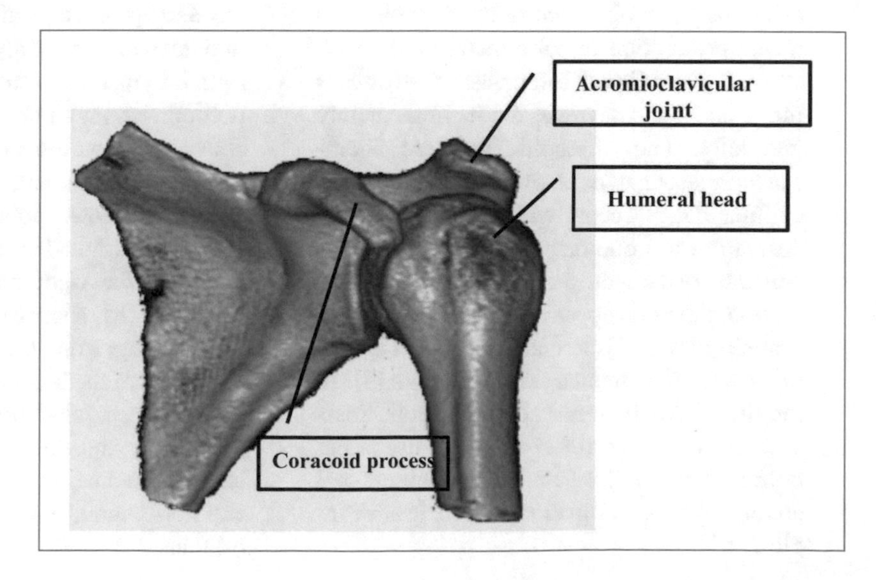

Fig. 2.2.1 Reconstruction of the humeral head and its articulation

G. Ausili Cefaro, D. Genovesi, C.A. Perez, *Delineating Organs at Risk in Radiation Therapy*,
DOI: 10.1007/978-88-470-5257-4_2-2, © Springer-Verlag Italia 2013

Table 2.2.1 Physiopathology of bone damage

Immature bone damage to the growing cartilage	Mature bone damage to osteoblasts
Within 2–4 days after exposure: chondrocytes in the temporary calcification zone swell, degenerate, and crash; consequently, their number decreases ↓ Histological recovery: high radiation exposure up to 12 Gy causes more serious cellular damage 1–2 months after the irradiation of the long bone: metaphyseal sclerosis ↓ Impairment and enlargement of the growth cartilage ↓ Return to normal within 6 months	Damage to osteoblasts ↓ Reduced production of bone matrix and increased resorption of osteoclasts

- Immature bone: Chondrocytes in the epiphyseal growth cartilage are the most radiation-sensitive areas of immature long bone. Microscopic changes in growth cartilage chondrocytes may appear following doses <3 Gy, and growth may slow down after just 4 Gy. Usually, histological recovery occurs up to 12 Gy, but higher levels of exposure may cause more serious cellular damage [5]. Alterations may appear 6 months or more after exposure, and it is not clear whether these are secondary to vascular or cellular damage or even a combination of the two [6].
- Adult bone: Irradiation causes damages to osteoblasts, with a consequent reduction in the production of bone matrix and an increased resorption of osteoclasts. Osteoblasts may undergo necrosis both immediately and later. The threshold for these alterations is estimated at around 30 Gy, and cellular death occurs with a dose of 5 Gy. Alteration to bone ranges from slight osteopenia to osteonecrosis. One year after irradiation, the bone appears osteopenic in radiographic views [3]. Radiation-induced alterations were first named actinic osteitis [7]. In the literature, the terms actinic osteonecrosis and osteoradionecrosis are also used, but both indicate major damage to the bone that enables the determination of cell death (Table 2.2.1).

2.2.2 Respiratory Apparatus

2.2.2.1 Bronchial Tubes

At the carina (located at the T4-T5 thoracic vertebrae), the trachea divides into two branches: the right and left main bronchi. The right bronchus is in a more vertical position compared with the left and forms a 20° angle with the longitudinal axis of the trachea. It is 2.5-mm long, with a diameter of 15 mm, and is the starting point for the following:

- right superior lobar bronchus, which runs laterally and upward and, after 10–15 mm, divides into three segments: apical, front, and rear;
- right intermediate or middle bronchus, which is continuous with the main lobar right bronchus, runs downward for 2 cm from the back wall of the distal part, where the apical bronchus of the lower lobar bronchus (Nelson's bronchus) originates; from the intermediate bronchus the right median lobar bronchus starts, which is about 6 mm in diameter and 10-25-mm long; from it, two segmental branches – lateral and medial – arise;
- right inferior lobar bronchus, which travels posteriorly and downward from the median lobar bronchus, is where the paracardiac, anterior, lateral, and posterior basal bronchi originate.

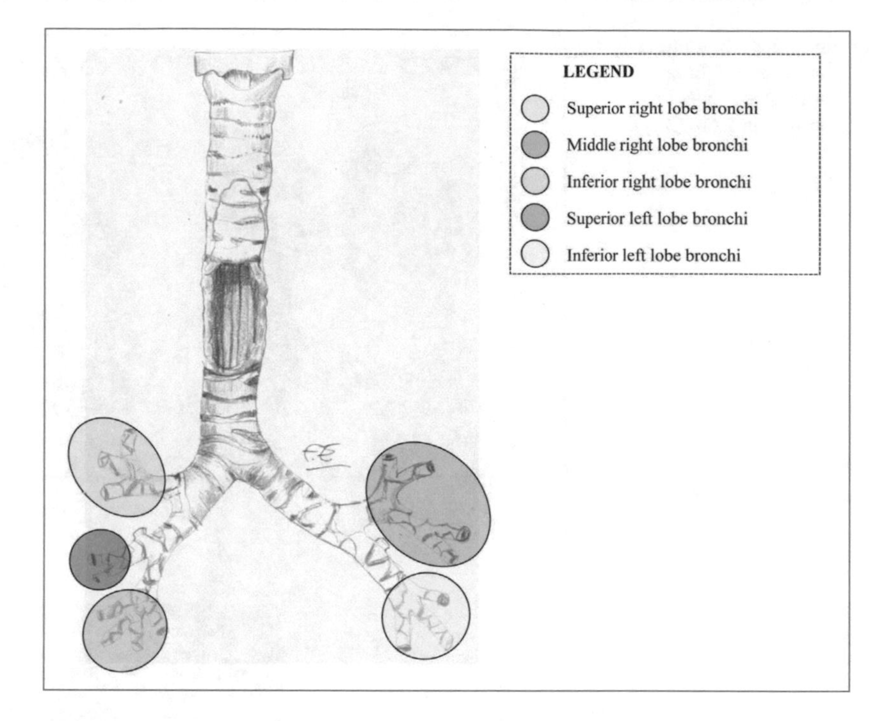

Fig. 2.2.2 Trachea and bronchi. Drawing by Franca Evangelista

The left bronchus forms a 40–50° angle with the longitudinal axis of the trachea; it is about 5-cm long and has a diameter of about 11 mm. It can be divided into the following:

- the left superior lobar bronchus, directed horizontally and forward, is divided into three segmental branches: apical, posterior, and anterior;
- the left lingular bronchus, which corresponds to the right middle lobar bronchus, is divided into upper and lower;
- the left inferior lobar bronchus, which is divided into the apical bronchus of the inferior lobe and the anterior, posterior, medial, and lateral basal bronchi [8] (Fig. 2.2.2).

2.2.2.2 Lungs

Lungs are situated in the pleura–pulmonary cavities and separated medially by the heart and mediastinal structures. The lungs are surrounded by the serosa, which is comprised of two layers (visceral pleura and parietal pleura) continuous to one another at the level of the hilum. Between the two layers is a space called the pleural space, which contains a thin, intrapleural liquid layer.

The surface of the lungs is crossed by scissures, which go deep into the hilum, dividing the organs into lobes: oblique scissure (on the left); oblique scissure (or principal) and horizontal scissure (on the right). In the lung, two vascular systems can be distinguished – pulmonary and bronchial:

- the pulmonary vessel system is functional, forming "the small circulation" and consisting of pulmonary arteries and veins;
- the bronchial vessel system is nutritional, forming "the big circulation" and consisting of bronchial arteries and veins (Fig. 2.2.3).

Ten areas (or segments) can be distinguished in the lung; each contains hundreds of independent entities, named pulmonary lobes, which are connected to each other by the interstitial connective tissue. The lobe structure is the following: intralobular bronchioles → terminal bronchioles → respiratory bronchioles → alveolar ducts → alveoli (alveoli are the primary site where gas exchange takes place). Their walls are composed of lining epithelium (alveolar epithelium) and an underlying connective layer rich in capillaries.

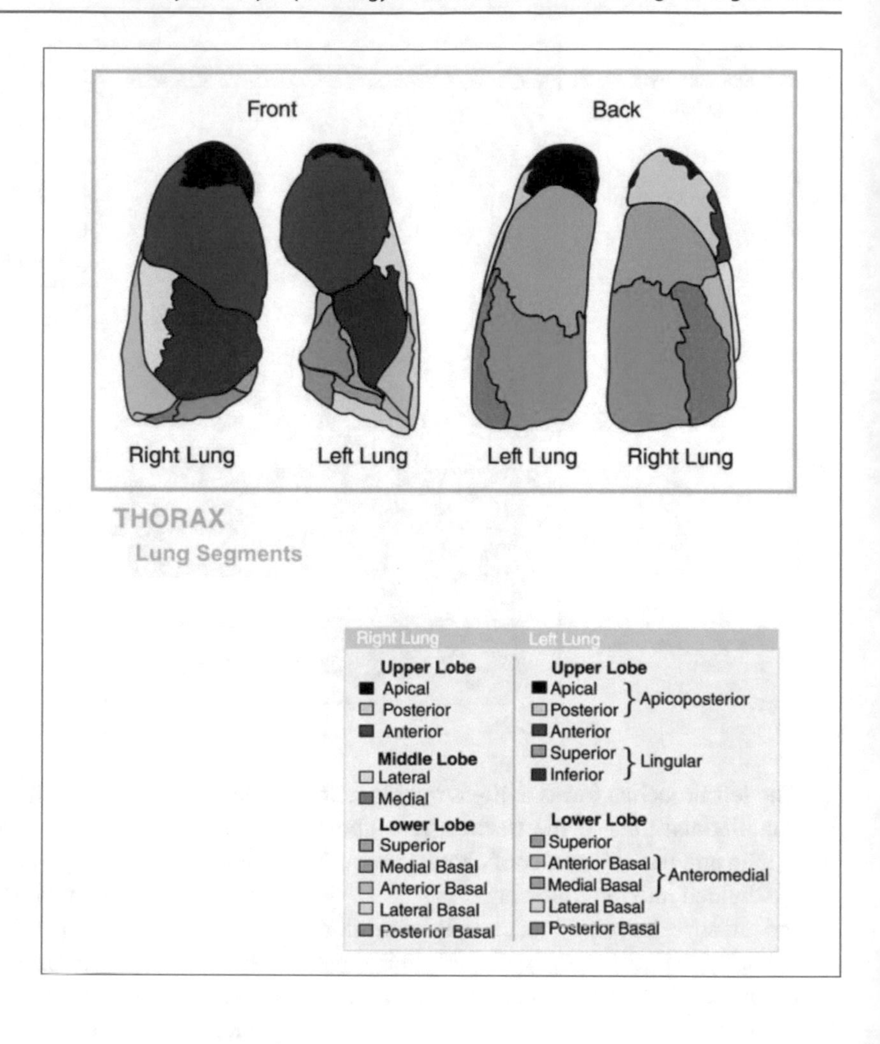

Fig. 2.2.3 Anatomical picture or the thorax. Reproduced with permission from [32]

Oxygen supply to tissue throughout the body, eliminating carbon dioxide, maintaining the correct acid-base balance in the blood, phonation, and protecting against pathogenic and irritating factors present in the air, are among the main functions of the respiratory system. Breathing, or lung ventilation, is the process that allows gas exchange. Inhalation and exhalation are determined by the difference between atmospheric and alveolar pressure, a pressure gradient that occurs when lung volume changes. Inhalation is determined by contraction of the diaphragm and external intercostal muscles that expand the thoracic cavity. These processes cause reduced intrapleural pressure (the lung-expanding factor), and alveolar pressure subsequently becomes lower than atmospheric pressure. During breathing at rest, exhalation occurs when the thoracic walls and lungs passively return to their original position. Active or forced inhalation involves contraction of the internal intercostal and abdominal muscles.

Radiation-induced lung diseases are quite common complications of radiotherapy to the thorax to treat mediastinal, mammary, and lung tumors. Depending on the time after radiotherapy course, these diseases may appear in the form of postactinic pneumonia (acute phase) or localized pulmonary fibrosis (late phase). The acute phase usually appears 4–12 weeks after the end of radiation treatment, even though it may appear as early as the first week, especially in cases of concomitant radiochemotherapy [9–11]. The late phase usually appears 6–24 months after the end of radiation treatment and it may remain stable even after 2 years [9–11]. In some patients, ac-

Table 2.2.2 Physiopathology of lung damage

Acute phase (4–12 weeks after the end of radiotherapy)	Intermediate phase (2–9 months after the end of radiotherapy)	Chronic phase (≥9 months after the end of radiotherapy)
Loss of type I pneumocytes ↓ Increased capillary permeability ↓ Interstitial and alveolar edema ↓ Infiltration of inflammatory cells into alveolar spaces	Obstruction of lung capillaries by platelets, fibrin, and collagen ↓ Infiltration of fibroblasts into the alveolar wall and interstitial fibrosis with collagen bands ↓ ↓ If slight damage If severe damage ↓ ↓ Alteration mitigation Consequent chronic phase	Progressive thickening of the alveolar ducts and increasing vascular sclerosis

tinic fibrosis may appear even without manifestation of a preceding acute pneumonia.

Actinic pneumonia is the manifestation of an acute reaction of the lung to radiation and presents as follows: loss of type I pneumocytes → increased capillary permeability with consequent interstitial or alveolar edema → input of inflammatory cells into alveolar spaces [12].

An intermediate phase (2–9 months after the end of radiotherapy) may occur, with obstruction of lung capillaries by platelets, fibrin, and collagen, with fibroblast infiltration into the alveolar wall and interstitial fibrosis with collagen bands. If the radiation-induced damage is slight, these alterations may mitigate; if the damage is severe, a chronic phase follows (usually ≥9 months after radiotherapy), which is dominated by progressive thickening of the alveolar ducts and increasing vascular sclerosis [12–14] (Table 2.2.2).

2.2.3 Heart

The heart is an unpaired, muscular, hollow organ situated between the lungs in the mediastinal space. It is contained in a fibroserous structure, the pericardium, which fixes it to the diaphragm and, at the same time, isolates it from nearby organs. The heart is divided into four parts: two superior (right and left atria) and two inferior (right and left ventricles). In the right atrium is the outflow of both the superior and inferior vena cava and the coronary sinus, where the venous blood flows from the heart walls. The right atrioventricular cavities are continuous through the right atrioventricular orifice (tricuspid orifice), which is protected by the tricuspid valve. In addition to the atrioventricular orifice, in the right ventricle is a pulmonary orifice (or ostium) containing a valvular apparatus composed of three semilunar valves, through which it is connected to the pulmonary trunk. The four pulmonary veins outflow into the left atrium – two for each side. The left atrioventricular cavities are continuous through the left atrioventricular orifice (or mitral ostium), in which the mitral valve (or bicuspid) is located. The mitral valve is composed of two cusps that enter the ventricular cavity. The left ventricle is connected to the aorta through the aortic orifice, which houses three aortic semilunar valves, the structure and insertion mode of which are similar to those of the pulmonary trunk.

The right and left coronary arteries supply the myocardium. The right coronary artery supplies the right ventricle and exits from the posterior and inferior walls of the left ventricle. After a first tract, called the common trunk, the left coronary artery is divided into two branches: the anterior descending artery, which runs along the anterior wall of heart, and the circumflex artery, which runs along the lateral wall of the left ventricle. Cardiac veins are tributaries of the coronary sinus, which flows into the right atrium [15] (Fig. 2.2.4).

Radiation-induced heart diseases (RIHD) represent clinical and pathological damage to

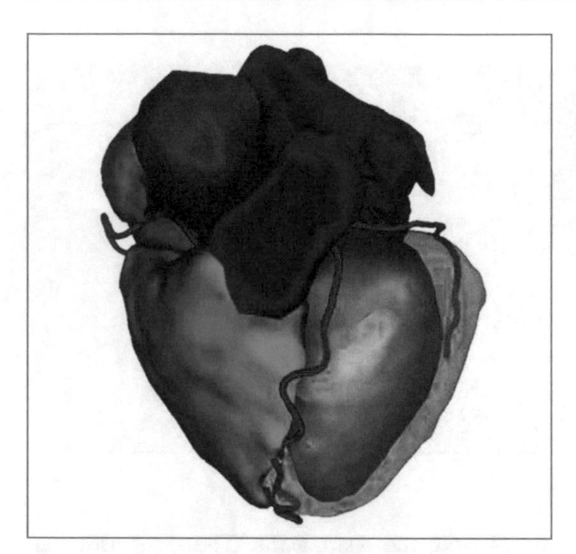

Fig. 2.2.4 Example of whole-heart and coronary-artery segmentation reconstruction. Reproduced with permission from [31]

the heart. Modern radiation techniques, like dose fractionation, and reduced irradiated volume of the heart when treating several neoplasms have reduced the frequency of RIHD over the last decade. As all heart structures may be susceptible to radiation-induced damage [16], the range of clinical manifestations is quite wide and consists of:

- acute pericarditis during course of treatment;
- late acute pericarditis;
- pericardial effusion;
- constrictive pericarditis;
- cardiomyopathy;
- valve damage;
- conduction abnormalities;
- ischemic coronary artery disease.

Cardiovascular toxicity manifestations may arise in patients irradiated because of lymphomas or pulmonary, mammary, or esophageal neoplasms; the damage entity is strictly connected

Table 2.2.3 Physiopathology of cardiac damage

Anatomical structure	Damage mechanism
Pericarditis	Collagen substitutes the normal adipose tissue ↓ Formation of fibrinous exudate on the surface and in the interstitium ↓ Proliferation of small blood vessels Later on, fibrinous exudates tend to organize ↓ Fibrotic evolution ↓ Onset of constrictive pericarditis
Cardiomyopathy	Interstitial fibrosis of pericellular and perivascular myocardium
Valvular affections	Fibrous scars on vascular flaps ↓ Stenosis and regurgitation
Conduction abnormalities	Myocardial fibrosis ↓ Conduction alteration or interruption at ventricular level or in the atrioventricular node
Coronary artery diseases and damage to the large vessels	Coronary arteries: Alterations of small vessels and capillaries at level of the endothelium ↓ Reduction of coronary arterial lumen ↓ Stenosis and thrombosis Large vessels: Alterations at the endothelial level ↓ Increased incidence of atherosclerotic degeneration and/or stenosis

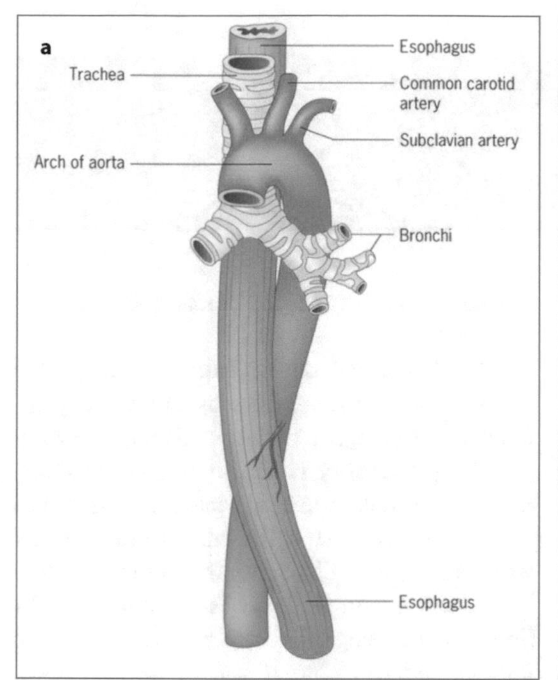

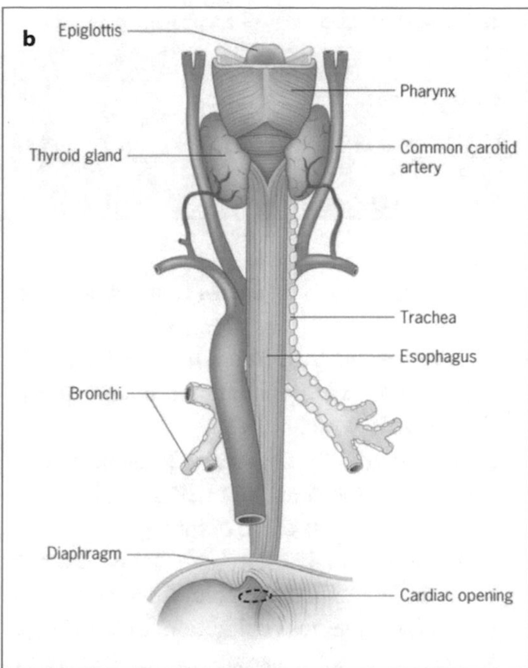

Fig. 2.2.5a,b Extension of the esophagus and its connections. The esophagus is a collapsed muscular tube that is approximately 25 cm in length. It has three constant constrictions. The first one is at the level of the cricoid cartilage; the second is at the level of the T4 vertebra, behind the bifurcation of the trachea (bronchial-aortic constriction); and the third is at the level of the T10 vertebra, where the esophagus passes through the esophageal hiatus of the diaphragm. There are two sphincters in the esophagus–the upper and the lower (or cardiac) esophageal sphincters. Reproduced with permission from [33]

with radiation dose. The most serious manifestations arise when the whole volume of the heart and pericardium are exposed to radiation. With conventional treatment, with doses of ~40 Gy for more than half of the heart volume, a maximum of 5 % of adult patients are likely to experience cardiac/pericardial disease. Age at the time of radiotherapy is an important risk factor as regards coronary artery damage; it is more evident in patients >50 years. Cardiac damage is much more frequent and serious when radiotherapy is concomitant or sequential with chemotherapy, particularly in cases in which cardiotoxic drugs are involved. RIHD can be described as being the result of micro- and macrovascular damage [17]. Damage to microcirculation begins with an alteration of the endothelial cells within the various cardiac structures. Swelling of capillaries and progressive obstruction of the vessel lumen cause ischemia that, in turn, leads to a substitution of cardiac tissue with fibrotic tissue [18]. Macrovascular damage results from damage to large vessels and causes of worsening in the formation of atherosclerotic lesions (Table 2.2.3).

2.2.4 Esophagus

The esophagus is the feeding canal that extends from the pharynx to the stomach and is approximately 25-26-cm long. From the outside in, the wall of esophagus consists of a tunica mucosa (stratified epithelium), a tunica submucosa, a muscular tunica (circular internal; longitudinal external), and a tunica adventitia. The esophagus is lacking of a serous tunica and is composed of cervical, thoracic, diaphragmatic, and abdominal portions (Fig. 2.2.5a,b):

- cervical esophagus, 4-5-cm long, is situated between the body of the cervical vertebra

Table 2.2.4 Physiopathology of esophageal damage

Acute phase	Late phase
Basal-cell necrosis ↓ Submucosal edema ↓ Microcirculation degeneration	Extended submucosal fibrosis with vascular alteration

C4 and the superior edge of thoracic vertebra T2;

- thoracic esophagus, 16-cm long, is located in the posterior mediastinum and extends up to the thoracic vertebra T11;
- diaphragmatic esophagus, 1-2-cm long, corresponds to the short tract that inserts into the esophageal hiatus of the diaphragm;
- abdominal esophagus, 3-cm long, extends from the hiatus to the cardia, the junction between the squamous epithelium of the esophagus and the columnar one of the stomach; it connects at the front with the lower surface of the left lobe of the liver and posteriorly with the abdominal aorta and medial pillars of the diaphragm; on the right, it is connected with the caudate lobe of the liver and on the left with the fundus of the stomach; the abdominal tract of the esophagus is covered with the peritoneum only on its anterior surface.

The function of swallowing implies a series of muscular contractions that push food from the oral cavity to the stomach. Acute esophagitis, which usually arises within 3 months after the beginning of radiotherapy, is quite common in patients treated for thoracic tumors. The frequency and severity of injuries depend on the total amount of radiation dose administered, on fractionation, and on any concomitant chemotherapy. Concomitant chemotherapy or altered fractionation may cause acute esophagitis severe enough to require hospitalization and interruption of radiotherapy in 15–25% of patients. Late injuries are less commonly registered because of a very low cancer-specific survival in many thoracic neoplasms. Standard or hypofractionation dose-escalation schemes may increase the risk of late esophageal toxicity in long-term survivors [19, 20]. Death due to late injuries (i.e., tracheoesophageal fistula or esophageal perforation) has been registered in only 0.4–1 % of patients [21, 22].

Dysphagia and odynophagia, linked to the onset of mucositis, are the most common symptoms of esophageal radiation damage. Symptomatology severity is usually slight and short-lived, even though in some patients it may last weeks or months after the end of radiotherapy with consequent difficulty in feeding. Sometimes, reflex pain in the chest may occur [23]. From an endoscopic point of view, acute actinic injuries and late injuries can be distinguished. In acute esophagitis, alterations that range from diffuse erythema with mucosal fragility to erosion, sometimes coinciding, and ulcerations can be observed. From a histological point of view, at this stage, basal cell necrosis, submucosal edema, and microcirculation degeneration are registered. Late actinic injuries (onset from 3 months to many years after radiotherapy) consist of more or less extended tubular stenosis covered with changes of mucosa, chronic ulcers, and tracheoesophageal fistulae (Table 2.2.4).

2.2.5 Spinal Cord

The brain and the spinal cord form the central nervous system (CNS), which collects, transmits, and integrates information. The spinal cord is contained in the vertebral canal and enveloped in the meningeal dural sac. It is in the shape of a cylinder, with an average diameter of 8–10 mm, and extends from the great occipital foramen to the first or second lumbar vertebra, therefore it does not entirely fill the vertebral canal. The lumbar portion of the vertebral canal contains roots which start from the region of the conus midollaris: the cauda equina. The spinal cord, just like

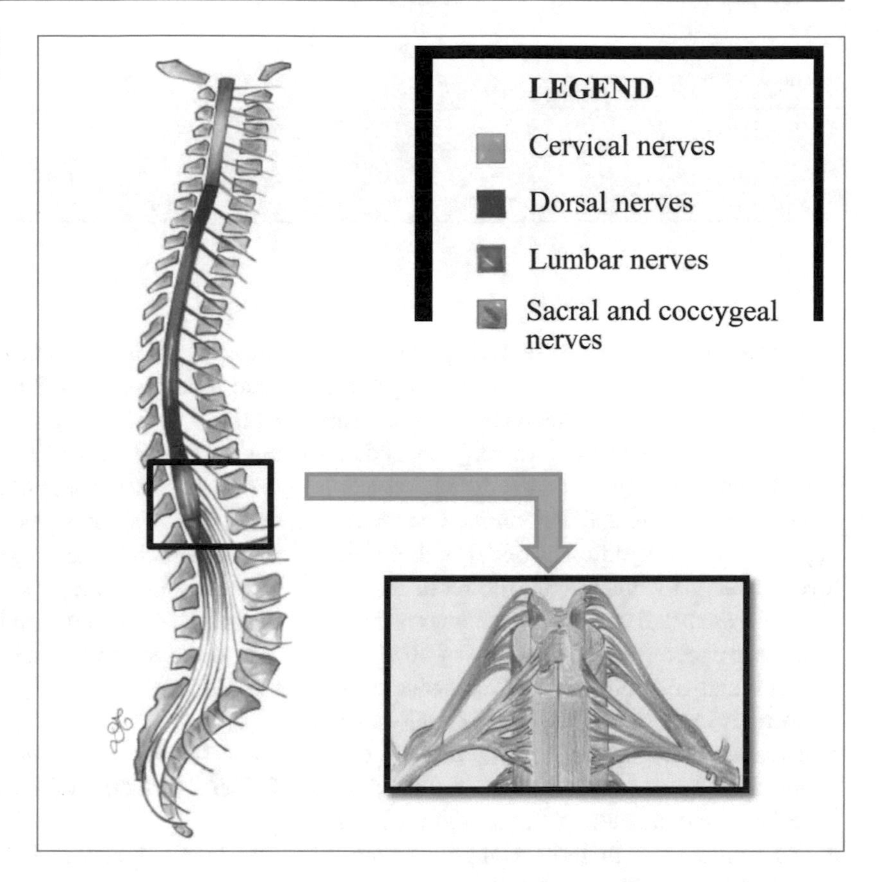

Fig. 2.2.6 Spinal cord and spinal nerves. Drawing by Danila Trignani

the spine, is divided into segments (cervical, thoracic, lumbar, sacral) that, in cross-section, are composed of grey matter inside and white matter outside:

- In the butterfly-shaped grey matter, two anterior, two posterior, and lateral horns can be distinguished. The two anterior horns innervate the skeletal muscles. Thus, the anterior (ventral) root is the motor root; it contains somatic motor fibers (voluntary) and visceral effectors. The posterior root is sensory and contains afferent fibers, somatic and visceral; there, deriving from the encephalic trunk, pain-suppression axons end. In the cells of the lateral horns, some orthosympathetic fibers (in the thoracic tract of the spinal cord) and parasympathetic fibers (at the level of the sacral spine) have their origin.
- The mass of white matter is, on the contrary, composed of long ascending and descending nerve bundles that connect the brain and the spinal cord, and are divided into two groups of three bundles each: with one group running on the right and one on the left, separated by the central spinal canal and posterior and anterior horns. Spinal nerves originate, one on the right and one on the left, in each medullary segment through the conjugate (or intervertebral) foramen; they represent the junction point of anterior and a posterior roots.

Spinal and cranial nerves, together with their relative ganglia, form the peripheral nervous system (PNS), which, through afferent nerve fibers conveys sensory information to the CNS and through efferent nerve fibers controls impulses from the CNS to organs and tissues [24]. Nerve-cell bodies are grouped in ganglia of the PNS and in nuclei of the spinal cord and brain, which form proper neuronal aggregations with specific characteristics as regards signal organization and elaboration. The PNS is divided into somatic (or voluntary) and autonomic (visceral or involuntary) nervous systems. Both are controlled and coordinated by the CNS (Fig. 2.2.6).

Table 2.2.5 Physiopathology of spinal cord damage

6–12 months after the end of radiotherapy	1 year or more after the end of radiotherapy
Spinal white-matter degeneration ↓ Demyelination (appears earlier, gradually, and progressively)	Primary vascular damage → necrosis (appears later, but more suddenly and dramatically)

The reduced tolerance of the spinal cord is considered one of the major dose-limiting factors in radiotherapy for the treatment of neoplasms situated near the neuraxis. Toxicity depends not only on the total dose but on the dose per fraction, tissue oxygenation, and concomitant chemotherapy. Spinal cord syndromes occur with different levels of severity and length; the more extended (≥10 cm) the irradiated tract, the higher is the risk of disease occurrence. With a dose of 40 Gy with conventional fractionation, the incidence of myelopathy is low (<0.2–2 %), even though isolated cases of myelopathy onset resulting from doses lower than those considered safe (30–35 Gy) have been documented in the literature. In fact, the alteration is due to the combination of several factors, some of which are difficult to evaluate and predict, whereas the risk of myelopathy becomes significant as a consequence of a total dose >50 Gy. The pathogenic mechanism is due to two main modes: (a) degeneration of the spinal white matter, which occurs earlier (after 6–12 months) and in a progressive, gradual manner; (b) necrosis as a consequence of primary vascular injury, which usually appears later (≥1 year) but occurs unexpectedly and more dramatically.

The onset of radiation-induced myelopathy may be insidious due to latency of even more than 5–6 months. Myelitis may appear early and in a transiently (transitory myelopathy), with paresthesias and electric-shock-type sensation to limbs caused by flexion and extension movements of the head (Lhermitte's sign); it does not affect motor fibers. Objective signs and symptoms of more serious situations may include spasticity, weakness, hemiparesis, and the less-frequent Brown-Séquard syndrome (ipsilateral paralysis, discriminatory sensitivity, contralateral hemisensory loss of pain and temperature sensitivity), incontinence, and occasional pain. Late myelopathy, which may appear anywhere from 20 to 36 months after irradiation, is serious and rarely reversible. It shows both motor and sensory involvement corresponding to transverse myelitis, the signs of which depend on the tract to which the damage occurred. Myelopathy is usually fatal. Injuries to the cervical segment of the spinal cord determine a more serious risk than those to the thoracic segment [25, 26] (Table 2.2.5).

2.2.6 Brachial Plexus

The brachial plexus is an important neuronal structure that supplies sensory and motor innervation to the upper limbs. It originates in the roots of C5–T1 and partially in C4 and T2. Composition and anatomical connections of the brachial plexus are described in Figs. 2.2.7 and 2.2.8.

Brachial plexus injuries may arise in patients who undergo radiotherapy in the axillary and supraclavicular regions. Such complications are now quite rare and are connected to disease recurrence rather than to radiation-induced neuropathy (postactinic plexopathy). The risk of plexopathy varies from 0 % to 5 % and is associated with total radiation dose amount, fractionation, volume of irradiated plexus, and concomitance with chemotherapy [27–29]. Neurological damage resulting from radiotherapy may be observed many months or years after the end of radiation treatment [30]. The exact physiopathology of plexopathies is still uncertain. It seems that vascular alterations, actinic fibrosis involving nerve structures, and the direct effect of radiation on Schwann cells play an important role (Table 2.2.6).

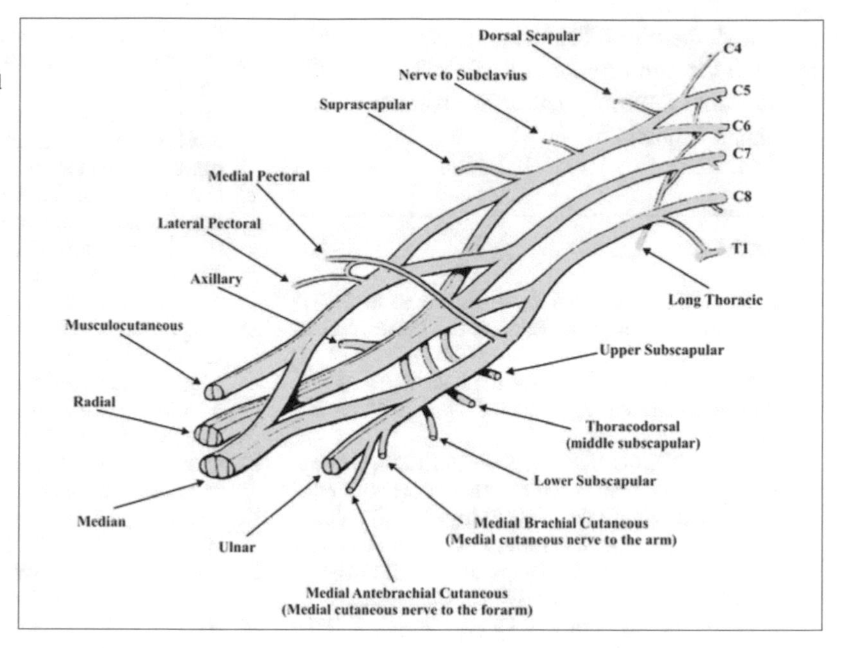

Fig. 2.2.7 Schematic representation of a prefixed brachial plexus. Reproduced with permission from [34]

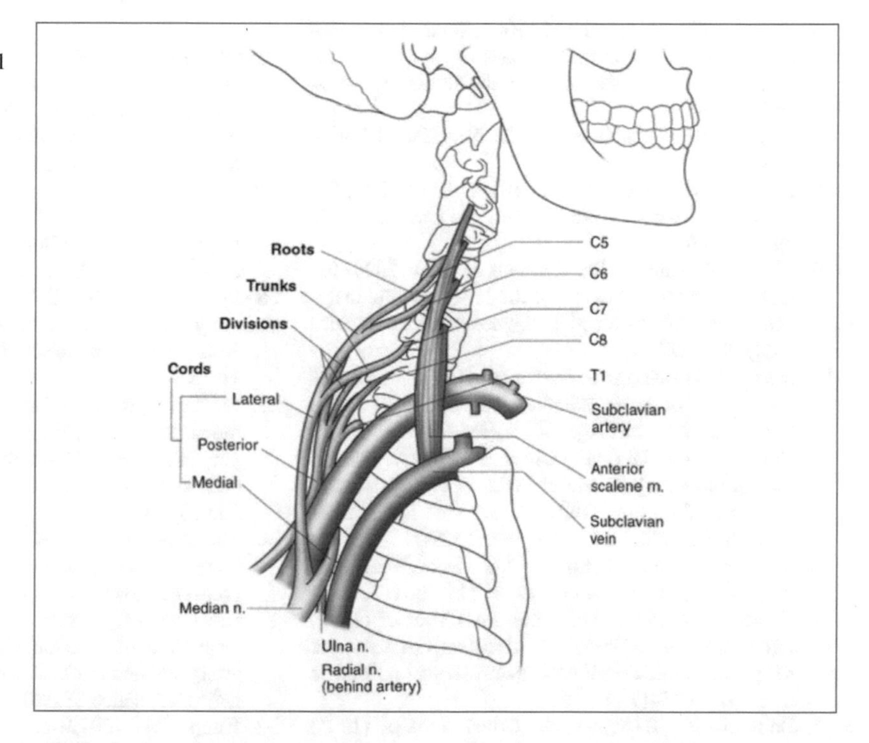

Fig. 2.2.8 Anatomical connections of the brachial plexus. Reproduced with permission from [35]

Table 2.2.6 Physiopathology of brachial plexus damage

Anatomical structure	Damage mechanism
Brachial plexus	Vascular alterations ↓ Actinic fibrosis involving nerve structures ↓ Direct effect on Schwann cells

Acknowledgement This chapter has been written with the contributions of Maria Taraborrelli, Lucia Anna Ursini, Monica Di Tommaso, and Marianna Trignani.

References

1. Balboni GC, Bastianini A, Brizzi E et al (1991) Anatomia Umana. Third Edition. Edi Ermes, Vol. 1, pp 207–209
2. Rubin P, Andrews JR, Swarm JR, Gump H (1959) Radiation-induced dysplasia of bone. AJR Am J Roentgenol 82:206–216
3. Howland WJ, Loeffler RK, Starchman DE, Johnson RB (1975) Postirradiation atrophic changes of bone and related complications. Radiology 117:677–685
4. Ergun H, Howland WJ (1980) Postradiation atrophy of mature bone. Crit Rev Diagn Imaging 12:225–243
5. Dalinka MK, Mazzeo VP (1985) Complications of radiation therapy. Crit Rev Diagn Imaging 23:235–267
6. Dalinka MK, Haygood TM (2002) Radiation changes. In: Resnick D (ed) Diagnosis of bone and joint disorders, 5th edn. Saunders, Philadelphia, pp 3393–3422
7. Ewing J (1926) Radiation osteitis. Acta Radiol 6:399–412
8. Balboni GC, Bastianini A, Brizzi E et al (1991) Anatomia Umana. Third Edition. Edi Ermes, Vol. 2, pp 319–346
9. Choi YW, Munden RF, Erasmus JJ et al (2004) Effects of radiation therapy on the lung: radiologic appearances and differential diagnosis. Radiographics 24(4):985–997
10. Ikezoe J, Takashima S, Morimoto S et al (1988) CT appearance of acute radiation-induced injury in the lung. AJR Am J Roentgenol 150(4):765–770
11. Collins J, Stern EJ (2007) Chest radiology, the essentials. Lippincott Williams & Wilkins, Philadelphia
12. Roswit B, White DC (1977) Severe radiation injuries of the lung. AJR Am J Roentegenol 129:127–136
13. Movsas B, Raffin TA, Epstein AH, Link CJ Jr. (1997) Pulmonary radiation injury. Chest 111;1061–1076
14. Charles HC, Baker ME, Hathorn JW et al (1990) Differentiation of radiation fibrosis from recurrent neoplasia: a role for 31^{P} MR spectroscopy? AJR Am J Roentgenol 154(1):67–68
15. Balboni GC, Bastianini A, Brizzi E et al (1991) Anatomia Umana. Third Edition. Edi Ermes, Vol. 1, pp 361–398
16. Adams MJ, Hardenbergh PH, Constine LS, Lipshultz SE (2003) Radiation-associated cardiovascular disease. Crit Rev Oncol Hematol 45:55–75
17. Corn BW, Trock BJ, Goodman RL (1990) Irradiation-related ischemic heart disease. J Clin Oncol 8:741–750
18. Seddon B, Cook A, Gothard L et al (2002) Detection of defects in myocardial perfusion imaging in patients with early breast cancer treated with radiotherapy. Radiother Oncol 64:53–63
19. Timmerman R, McGarry R, Yiannoutsos C et al (2006) Excessive toxicity when treating central tumors in a phase II study of stereotactic body radiation therapy for medically inoperable early-stage lung cancers. J Clin Oncol 24:4833–4839
20. Onishi H, Shirato H, Nagata Y et al (2007) Hypofractionated stereotactic radiotherapy (HypoFXSRT) for stage I non-small-cell lung cancer: Updated results of 257 patients in a Japanese multi-institutional study. J Thoracic Oncol 2(7 Suppl 3):S94–S100
21. Singh AK, Lockett MA, Bradley JD (2003) Predictors of radiation-induced esophageal toxicity in patients with non-small-cell lung cancer treated with three-dimensional conformal radiotherapy. Int J Radiat Oncol Biol Phys 55:337–341
22. Qiao W-B, Zhao Y-H, Zhao Y-B et al (2005) Clinical and dosimetric factors of radiation-induced esophageal injury: Radiation-induced esophageal toxicity. World J Gastroenterol 11:2626–2629
23. Rubin P, Casarett GW (1968) Clinical radiation pathology. W.B. Saunders Company, Philadelphia
24. Balboni GC, Bastianini A, Brizzi E et al (1991) Anatomia Umana. Third Edition. Edi Ermes, Vol. 3, pp 28-49
25. Okada S, Okeda R (2011) Pathology of radiation myelopathy. Neuropathology 21:247-65
26. Wong CS, Van Dyk J, Milosevic M et al (1994) Radiation myelopathy following single courses of radiotherapy and retreatment. Int J Radiat Oncol Biol Phys 30:575-81
27. Hardenbergh PH, Bentel GC, Prosnitz LR et al (1999) Postmastectomy radiotherapy: toxicities and techniques to reduce them. Semin Radiat Oncol 9:259–268
28. Recht A, Edge SB, Solin LJ et al (2001) Postmastectomy radiotherapy: clinical practice guidelines of the American Society of Clinical Oncology. J Clin Oncol 19:1539–1569
29. Pierce SM, Recht A, Lingos T et al (1992) Long-term radiation complications following conservative surgery (CS) and radiation therapy (RT) in patients with early stage breast cancer. Int J Radiat Oncol Biol Phys 23:915–923
30. Moore NR, Dixon AK, Wheeler TK et al (1990) Axillary fibrosis or recurrent tumor: an MRI study in breast cancer. Clin Radiol 42:42–46
31. Breeuwer M, Ermes P, Gerber B (2009) Clinical evaluation of automatic whole-heart and coronary-artery segmentation. Journal of Cardiovascular Magnetic Resonance 11:220
32. Kitapci MT (2012) Atlas of Sectional Radiological Anatomy for PET/CT. Springer
33. Najam A, Ajani J, Markman M (2003) Atlas of Cancer. Volume 4, Chapter 31
34. Pellerin Megan, Kimball Z, Tubbs RS, Nguyen S, Matusz P, Cohen-Gadol AA, Loukas M (2010) The prefixed and postfixed brachial plexus: a review with surgical implications. Surgical and Radiologic Anatomy. 32:251-260
35. Saifuddin A (2003) Imaging tumours of the brachial plexus. Skeletal Radiology 32:375-387

Chapter 2.3
Abdomen

2.3.1 Liver

The liver, the most voluminous gland in the human body, is located in the superior portion of the abdominal cavity and occupies the entire right hypochondrium, part of the epigastrium, and most of the cranial area of the left hypochondrium. The anterosuperior part is divided by the superior sagittal groove into the right and left lobes. The inferior part is divided into four lobes, right, left, quadrate, and caudate (Spigelian lobe), and is crossed by three grooves, right sagittal, left, and transverse (hepatic hilum). Liver vascularization occurs through the hilum, which receives the hepatic artery (branch of the celiac trunk) and portal vein, the latter being created by a confluence of veins that drain blood from the abdominal organs (stomach, small and large bowel, spleen, pancreas). Hepatic veins transport liver deoxygenated blood and blood which is filtered by the liver to the inferior vena cava. They originate in the liver lobule's central veins (Fig. 2.3.1).

The liver performs innumerable functions: storage of metabolites (e.g., glucose, iron, vitamins), blood depuration, regulation of protein, lipid, and carbohydrate metabolism, breakdown of nutrients and drugs in blood carried from the digestive system for easier use by the rest of the body, and regulation of blood coagulation [1, 2].

Hepatic damage from ionizing radiation changes according to the dose delivered to the entire organ, a segment, or a lobe. In radiation hepatitis, acute hepatic damage is commonly reported [3].

The first radiation damage are visible at a dose of 15 Gy on the entire organ; irradiation at 45–50 Gy with conventional fractionation on a limited area of the liver does not produce appreciable consequences from a clinical point of view. Radiation-induced damage at the hepatic level is initially represented by hepatomegaly (in the absence of jaundice), ascites, and an increase in hepatic functionality values (mainly of alkaline phosphatase and glutamic oxaloacetic transaminase with a decrease in bilirubin values). Radiation-induced liver disease is characterized by occlusion and obliteration of hepatic lobule's

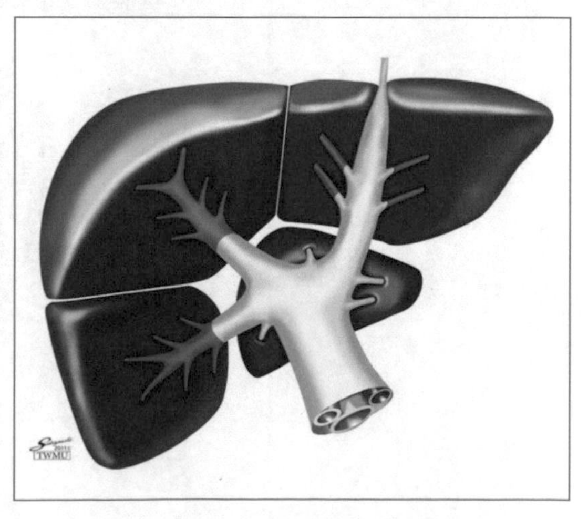

Fig. 2.3.1 The liver is divided into 3 segments and a caudate area according to the ramification of the Glissonian pedicles. Reproduced with permission from [17]

G. Ausili Cefaro, D. Genovesi, C.A. Perez, *Delineating Organs at Risk in Radiation Therapy*,
DOI: 10.1007/978-88-470-5257-4_2-3, © Springer-Verlag Italia 2013

Table 2.3.1 Physiopathology of liver damage

Anatomical structure	Damage mechanism
Endothelial cells	Activation of soluble coagulation factors → fibrin deposit → central veins occlusion → retrograde hepatocyte congestion → necrosis and increase in hepatic transaminase

central veins with retrograde congestion and subsequent secondary hepatocyte necrosis appearing between 2 weeks and 3 months after the end of treatment [4–6]. In most cases, the liver gradually repairs after subacute radiation damage. The architecture of all lobules is often distorted. Late lesions are generally asymptomatic, but sometimes they can be associated with clinical manifestations, defined as chronic hepatitis due to radiation (Table 2.3.1)

2.3.2 Kidneys

The kidneys are paired organs located in the posterosuperior abdominal cavity on either side of the spinal cord in the retroperitoneal space. Considering their relation with the vertebral column, the kidneys extend from the inferior margin of the 11th thoracic vertebra up to the superior margin of the third lumbar vertebra. The right kidney is lower than the left by about 2 cm due to its relation with the liver, which pushes it to a lower position.

The kidney is divided into an anterior part, a posterior part, a superior pole, an inferior pole, a lateral margin, and a medial margin. The latter is hollow medially, where a vertical incisure of about 3–4 cm is present – the renal hilum – which allows lymphatic and blood vessels, as well as nerves and the renal pelvis, to pass. The renal lodge is delimited by a connective fascia, which, being near the kidney, gets thicker, forming the renal fascia.

The kidney consists of two areas: the peripheral, or cortical, which surrounds the central area, the medulla (Fig. 2.3.2). It performs several functions, such as excreting the products of catabo-

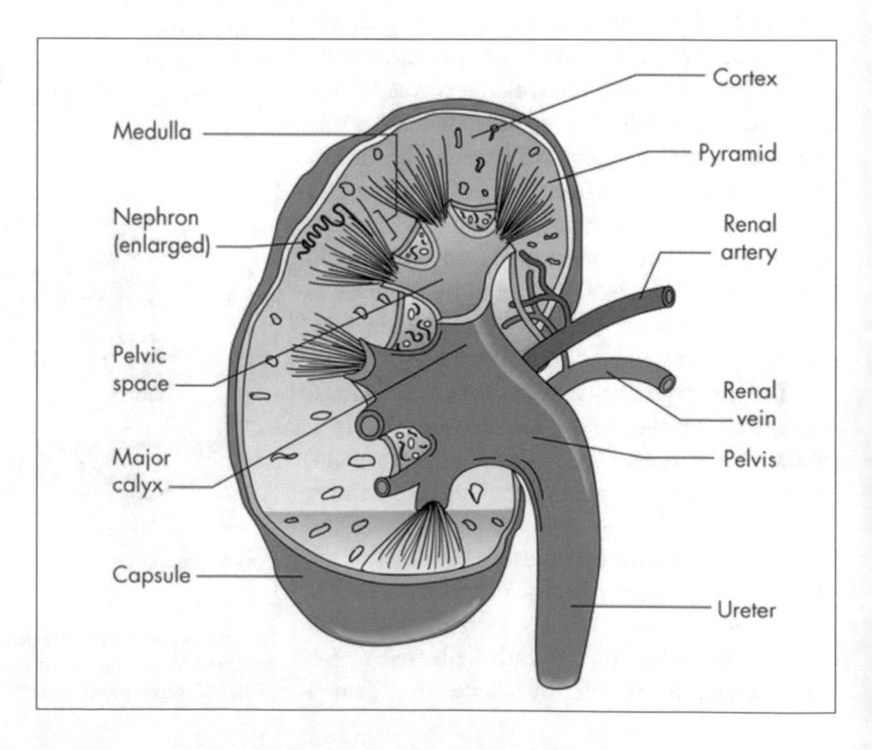

Fig. 2.3.2 Kidney anatomy. Reproduced with permission from [18]

Table 2.3.2 Clinical syndromes in relation to latency

Syndrome	Latency (months)
Acute nephritis	6–12
Chronic kidney disease	>18
Benign hypertension	>18
Malignant hypertension	12–18

Table 2.3.3 Physiopathology of renal damage

Anatomical structure	Damage mechanism
Vascular elements and tubular epithelium	Intimal necrosis → subendothelium fibrous thickening → collagen replacement → glomerular sclerosis (acute focal damage) Progressive proliferation of fibrous phenomena → reduction of parenchyma component → organ atrophy (late generalized damage)

lism and external substances and regulating water and electrolyte balance, body-fluid osmolarity, acid-base balance, erythropoiesis, and blood pressure [7].

The physiopathology of renal damage due to radiation is not yet well understood. The acute effect of radiation manifests as a change in glomerular and juxtaglomerular cells of the capillary endothelium, which detach from basal membranes and cause occlusive phenomena in both interlobular arteries and efferent arterioles. Afterwards, an increase in capillary permeability and interstitial edema follow. During the first days, the irradiated kidney shows modest vessel congestion associated to expanded parenchymal edema.

Changes in tubular epithelium are evident 15–20 days from the beginning of radiation therapy. In the subacute period (from 6 to 12 months after radiotherapy), the following phenomena are to be observed from an anatomical and pathological perspective: intimal necrosis, fibrous subendothelial thickening, tubular atrophy, collagen replacement, and glomerular sclerosis. Late damage occurs from 1 to 5 years, and it is represented by phenomena associated with a reduction of the parenchymal component, resulting in atrophy and sclerosis of the organ. In later phases, the predominant damage is caused by mesangial cells progressing toward hypertrophy and hypoplasia, with an increase in the matrix, which is the cause of the capillary obliteration that results in nephrosclerosis [8] (Tables 2.3.2 and 2.3.3).

2.3.3 Stomach

The stomach, situated between the esophagus and small intestine, lies in the abdominal cavity under the diaphragm and occupies the left hypochondrium and part of epigastrium. The stomach is composed of two walls, anterior and posterior, delimited by two margins, a small curve on the right and a large curve on the left. The small curve starts from the cardia, continues to the right margin of the esophagus, descends almost vertically, bending on top and posteriorly (angulus), finally reaching the pylorus where it continues with the superior margin of the duodenum. The large curve starts from the superior cardia contour in the cranial direction, forms with the left margin of the esophagus at an acute angle, descends until reaching the pylorus, and continues with the inferior margin of the duodenum.

The stomach relates superiorly to the esophagus through the cardia and inferiorly to the duodenum through the pylorus. Three general sections can be distinguished, represented in the craniocaudal direction by: fundus, corpus, and pylorus. The fundus is the most cranial portion, adapting to the hollow space of the diaphragm; the limit between fundus and corpus is conventionally defined by a horizontal plane passing through the cardia; the limit with the pylorus is determined by an oblique line reaching from the angulus to the large curve. The pylorus joins with the corpus at a 90° angle, creating a

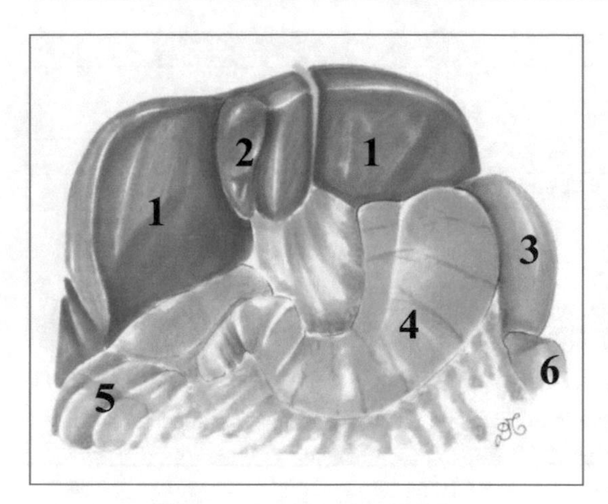

Fig. 2.3.3 Anatomical picture of stomach. *1* Liver, *2* gallbladder, *3* spleen, *4* stomach, *5* ascending colon, *6* descending colon. Drawing by Danila Trignani

incisure, the angular fold, on the small curve (Fig. 2.3.3).

Stomach vascularization is provided by the branches of celiac trunk: the left gastric artery originates directly from the celiac trunk; the right gastric artery and gastroduodenal artery originate from the common hepatic artery; the left gastro-epiploic artery and short gastric vessels originate from the splenic artery [9] (Fig. 2.3.3).

The stomach performs motion and secretion functions that mix food with gastric juices, transforms it into chyme (a semifluid food resulting from digestion). It also performs gastric emptying, thus completing the gastric digestive phase [9].

The stomach is considered a serial organ consisting of a chain of functional units. Therefore necrosis of 5–25% can be considered lethal [10].

The physiopathological mechanism of actinic acute and chronic damage is well defined [11]. Actinic gastritis might begin as early as a week after radiotherapy begins and can be persistent at different levels for more than one month. Irradiation causes an early block of mitotic activity in the most radiosensitive areas, as mucosa in the gastric fundus. Degenerative modifications occur later in other areas of the stomach, where the superficial epithelium, parietal cells, and zymogene cells show minor radiosensitivity. The pylorus and glands of the antral area show even lower sensitivity. Alterations induced by ionizing radiations both on mucosa and glands include edema, vasodilatation, and stromal inflammatory signs; subsequently, occlusion processes occur at the vascular level. Following these alterations, mucosa thickness is reduced, the glandular system becomes atrophied, and ulcerating processes might occur [12] (Table 2.3.4).

2.3.4 Small Bowel

The small bowel originates in correspondence with the pyloric sphincter, terminating at the level of the ileocecal valve. It expands, forming several inflexures, from the epigastrium to the right iliac fossa, thus occupying the major part of the abdominal cavity and descending into the small pelvis.

The duodenum is the first portion of the small bowel. It is about 30-cm long and originates at the first lumbar vertebra on the right of the median line, ending on the left of the second lumbar vertebra.

The mesenteric small bowel is the longest intestine, extending from the duodenum to the large bowel. It originates in correspondence to the second lumbar vertebra, terminating into the right iliac fossa, and continues into the large in-

Table 2.3.4 Physiopathology of gastric damage

Anatomical structure	Damage mechanism
Vascular system	Inflammation → edema with vasodilatation → vessel occlusion → progressive obliterating endoarteritis
Epithelium cells	Mitotic activity block → cell atrophy and death → reduced mucosal thickness → gland atrophy → ulcers

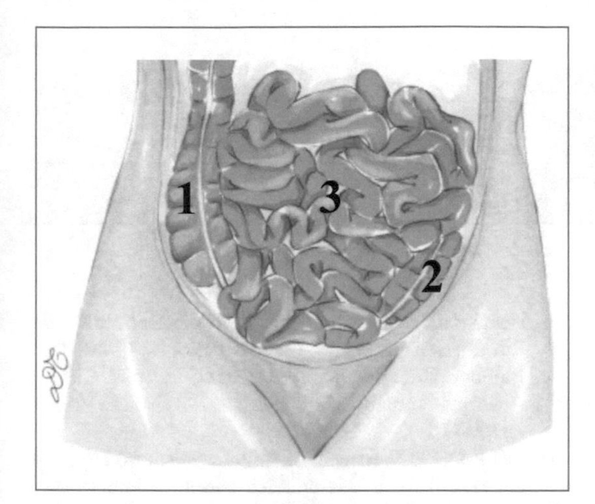

Fig. 2.3.4 Anatomical picture of the small bowel. *1* Ascending colon, *2* descending colon, *3* small bowel. Drawing by Danila Trignani

testine via the ileocecal valve. Two portions are to be distinguished: the jejunum and ileum. There is no real boundary between these two segments, although the jejunum is characterized by a relatively larger lumen, a thicker wall, and is more richly provided with villi and glands than ileum. This part of the small bowel is called mesenteric because it is compressed into the thickness of the free margin of a large peritoneal fold, the mesentery, detaching with its radix from the posterior abdominal wall (Fig. 2.3.4).

Vascularization is provided by the superior mesenteric artery through several collateral branches and their anastomosis [13].

The small bowel performs four main functions: food transport, nutrient absorption, motility, and fecal material containment (the latter is guaranteed particularly by the continence of the ileocecal valve). It also performs important endocrine and immune functions [13].

Small-bowel tolerance to ionizing radiation is one of the dose-limiting factors in abdominal–pelvic radiotherapy. The small bowel can be considered an organ with a significant correlation between irradiated volume and the probability of acute toxicity, independent of the dose received. Actinic enteritis shows a wide incidence range, being correlated with different factors: technical (total dose, irradiation volume, and fractionation) and clinical (age, general condition, hypertension, pelvic inflammatory disease, vascular disease, or diabetes) [14]. Actinic enteritis manifests as diarrhea and is variably associated with painful abdominal cramps, nausea, vomiting (rare), lack of appetite, and weight loss. Histological changes due to acute enteritis are particularly evident at the level of the mucosa.

Apoptosis of cryptic cells occurs, determining a reduction in villi as well as of mucosa thickening, with subsequent inflammatory processes until crypt abscess formation. The damage, evident in the third week of radiotherapy, tends to recover spontaneously after 3–4 days from the end of radiotherapy course, to be completed within 12–14 days [15]. In the case of actinic chronic enteritis, the histological pattern is modified in all components of the intestinal wall: mucosal atrophy occurs, with superficial linear ulcerations tending to confluence; the submucosa is the site of secondary thickening due to hyaline substance deposits; the muscular tunica alternates atrophic tracts with fibrotic tracts. The vascular pattern is characterized by secondary hypoafflux and obliterating endoarteritis, which is associated with changes in lymphatic circulation.

In the chronic condition, two peculiar clinical patterns can be distinguished: enteric toxicity of the medical type and chronic enteric toxicity of the surgical type. Globally, their incidence approximately varies between 10% and 50%, respectively [15]. Manifestation indicates a serious prognosis, with total survival rates at 5 years of 50–60%. In fact, severe chronic actinic enteritis manifestation indicates a survival rate four times lower than that expected for tumors requiring pelvic radiation treatment [16]. Chronic medical enteritis (functional damage), with a latency time of approximately 12–14 months, is caused by functional mucosal insufficiency. It is characterized by chronic diarrhea and malabsorption syndrome accompanied by steatorrhea due to pancreatic insufficiency, biliary salts, and fatty-acid malabsorption, with spasm-like abdominal pains.

Chronic surgical enteritis (anatomical damage) has a long latency time (ranging between 1 and 10 years) and is caused by deep mucosal ul-

Table 2.3.5 Physiopathology of small-bowel damage

Anatomical structure	Damage mechanism
Vascular system	Inflammation → edema with vasodilatation → vessel occlusion → obliterating progressive endoarteritis
Epithelium cells	Mitotic activity block → atrophy and cells death → reduced thickness of mucosa → ulcers

cerations, such as perforation, fistula, and hemorrhage. It therefore evolves into intestinal stenosis with bowel obstruction (Table 2.3.5).

Acknowledgement This chapter has been written with the contributions of Antonietta Augurio and Marianna Trignani.

References

1. Balboni GC, Bastianini A, Brizzi E et al (1991) Anatomia Umana. Third Edition. Edi Ermes, Vol. 2, pp 176–212
2. Guyton AC (1991) Trattato di fisiologia medica. Eighth Edition. Piccin, pp 842-847
3. Fajardo LF (1982) Pathology of radiation injury. Masson, New York
4. Wharton JT, Delclos L, Gallager S, Smith JP (1973) Radiation hepatitis induced by abdominal irradiation with the cobalt 60 moving strip technique. Am J Roentgenol 117:73–80
5. Ingold DK, Reed GB, Kaplan HS, Bagshaw MA (1965) Radiation hepatitis. Am J Roentgenol Radium Ther Nucl Med 93:200–208
6. Lawrence TS, Robertson JM, Anscher MS et al (1995) Hepatic toxicity resulting from cancer treatment. Int J Radiat Oncol Biol Phys 31:1237–1248
7. Balboni GC, Bastianini A, Brizzi E et al (1991) Anatomia Umana. Third Edition. Edi Ermes, Vol. 2, pp 355–389
8. Beauvois S, Van Houtte P (1997) Effets tardifs de l'irradiation sur le rein. Cancer Radiother 1:760–763
9. Balboni GC, Bastianini A, Brizzi E et al (1991) Anatomia Umana. Third Edition. Edi Ermes, Vol. 2, pp 92–115
10. Brick I (1995) Effects of million volt irradiation on the gastrointestinal tract. Arch Intern Med 96:26–31
11. Goldgraber MD, Rubin CE, Palmer WL et al (1954) The early gastric response to irradiation; a serial biopsy study. Gastroenterology 27:1–20
12. Coia LR, Myerson RJ, Tepper JE (1995) Late effects of radiation therapy on the gastrointestinal tract. Int J Radiat Oncol Biol Phys 31:1213–1236
13. Balboni GC, Bastianini A, Brizzi E et al (1991) Anatomia Umana. Third Edition. Edi Ermes, Vol. 2, pp 116–144
14. Letschert JG, Lebesque J, Aleman JV et al (1994) The volume effect in radiation-related late small bowel complications: results of a clinical study of the EORTC Radiotherapy Cooperative Group in patients treated for rectal carcinoma. Radiother Oncol 32:116–123
15. Polico C, Capirci C, Stevanin C et al (1993) Enteropatia da raggi. Scientifiche Nutricia, Milano, pp 1–44
16. Harling H, Balslev I (1988) Long-term prognosis of patients with severe radiation enteritis. Am J Surg 155:517–519
17. Yamamoto M, Katagiri S, Ariizumi S, Kotera Y, Takahashi Y (2011) Glissonean pedicle transection method for liver surgery (with video). Journal of Hepato-Biliary-Pancreatic Sciences 19:3-8
18. Malhotra V, Muravchick S, Miller R (2002) Atlas of Anesthesia. Volume 5, Chapter 8. Elsevier Health Sciences

Chapter 2.4
Pelvis

2.4.1 Rectum

The rectum, the end of the large bowel, is about 15-cm long, and consists of pelvic and perineal portions. The anatomic limit between the two portions is the intersection of the anus levator muscle. At its origin, the rectum generally corresponds to the third sacral vertebra, descending on the anterior side of the sacrum and coccyx and forming an anterior hollow curve (sacral curve) that is converted into a convex curve in correspondence with the prostatic apex in males and mid-vagina in females. The peritoneum partially covers the rectum; therefore, a peritoneal section and an extra-peritoneal section can be distinguished (Fig. 2.4.1).

Pelvic and perineal rectal relations are obviously different between males and females. In males, the pelvic rectum is related to the intestinal loops (peritoneal portion), bladder, posterior side of the prostate, deferent ducts, and seminal vesicles (subperitoneal portion). In females, it is related to intestinal loops (peritoneal portion) and posterior wall of the vagina, with interposition of the rectovaginal septum (subperitoneal portion). On the posterior side, it is related to the last three sacral vertebrae and coccyx and levator ani, piriform, and coccygeus muscles. In males, the perineal rectum is related to the prostate apex, membranous urethra, and urethral bulb, and in females to the posterior wall of the vagina.

The structure of rectal wall consists of three layers: mucosa, submucosa, and muscular [1] (Fig. 2.4.1).

Knowledge of the structural anatomy of the rectal wall is essential to understand the physiology and complex physiopathology of rectal damage following the end of radiotherapy course [2]. The rectum is a mixed-structure organ at risk, which is primarily composed of parallel subunits but also serial subunits; its tolerance to radiation depends on the dose that a certain percentage of its volume receives, as well as on the maximal dose at a specific point.

Radiotherapy can trigger both acute and late toxicities to the rectum. In the acute phase, it reduces the number of crypts and causes inflammatory cell infiltration into the rectal wall [3]. There is, however, no clear difference between symptoms of acute and late toxicity, and it is not clear when resolution of acute damage occurs and late damage begins. On the other hand, it is not yet clear whether symptoms and their related acute damage may coexist with symptoms of late damage. It has been demonstrated that acute toxicity predicts late toxicity (bleeding, proctitis, mucorrhea, incontinence, urgency, tenesmus), although factors related to the patient and previous surgery may affect rectal tolerance independently.

Fecal incontinence is a possible consequence of radiotherapy that, acting on sphincters, can determine a variation in sphincter pressure at rest and under effort 4–6 weeks following radiotherapy and probably persisting for 2 years after the

G. Ausili Cefaro, D. Genovesi, C.A. Perez, *Delineating Organs at Risk in Radiation Therapy*,
DOI: 10.1007/978-88-470-5257-4_2-4, © Springer-Verlag Italia 2013

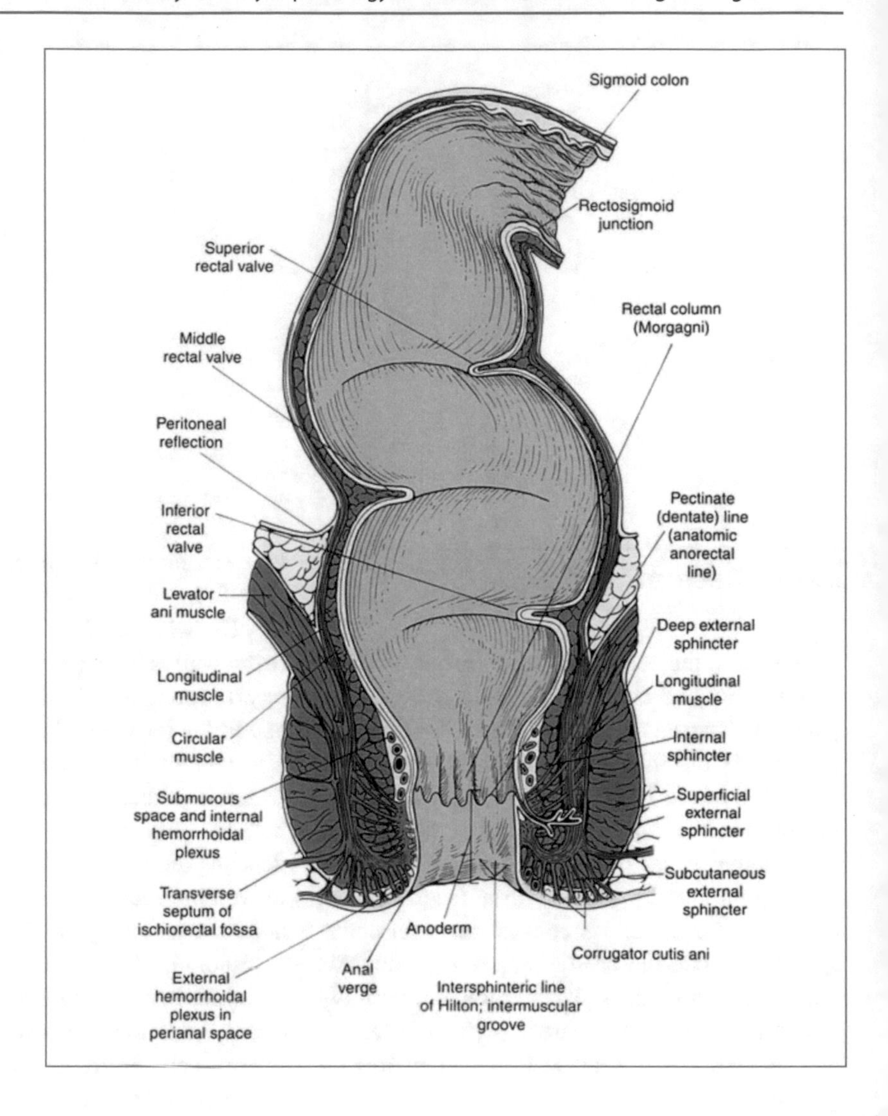

Fig. 2.4.1 Anatomical image of the rectum and anus. Reproduced with permission from [45]

end of radiotherapy. Reduced rectal volume at the border level of stimulus perception might suggest possible damage at the neuronal level, associated with myenteric plexus hypertrophy. Finally, fibrosis combined with a reduction of elastic fibers and an overall increase in sphincter rigidity might be associated with reduced rectal capacity [4, 5].

No morphological modifications occur at the level of the inner sphincter and Meissner plexus suggesting that damage following radiotherapy occurs primarily in the most external layers of the rectal wall [6, 7] (Table 2.4.1).

Table 2.4.1 Physiopathology of rectal damage

Anatomical structure	Damage mechanism
Rectal wall	Crypt reduction and inflammatory-cell infiltration (acute)
Muscular tunicae External anal sphincter	Elastic fiber reduction → collagen fibers deposit Rigidity and reduction of rectal compliance
Myenteric plexus	Hypertrophy → rigidity and reduction of rectum compliance

2.4.2 Bladder

The bladder is a hollow, muscular, and membranous organ in which shape, size, and placement change according to its filling and emptying status. The empty bladder is completely contained in the small pelvis, behind the pubic symphysis, in front of the uterus in females and the rectum in males. The superior portion is hollow and covered by the peritoneum, whereas the inferior portion, positioned on the posterior surface of the pubic symphysis, is convex. During filling, the bladder walls expand, and the superior part, the most extendible, rises and becomes convex, contributing to the egg-shaped aspect of the urine-filled bladder. When full, the bladder overlaps the superior margin, reaching the hypogastric region adjacent to the anterior abdominal wall. The expanded bladder can be described as follows: the base, or lower portion; the corpus (with a posterior, an anterior, and two lateral portions); and the apex, all showing different anatomical relations.

In men, the base relates to the prostate, seminal vesicles, and rectum (the latter with the interposition of the bladder–rectum fascia); in females, it relates to the upper third of the vagina (bladder–vaginal septum), the vaginal fornix, and the supravaginal portion of uterine neck. The corpus relates frontally to the pubic symphysis and internal obturator muscles and laterally to the side walls of the small pelvis. In females, it relates to the uterus on the posterior portion and in males to the pelvic colon and small bowel (Fig. 2.4.2).

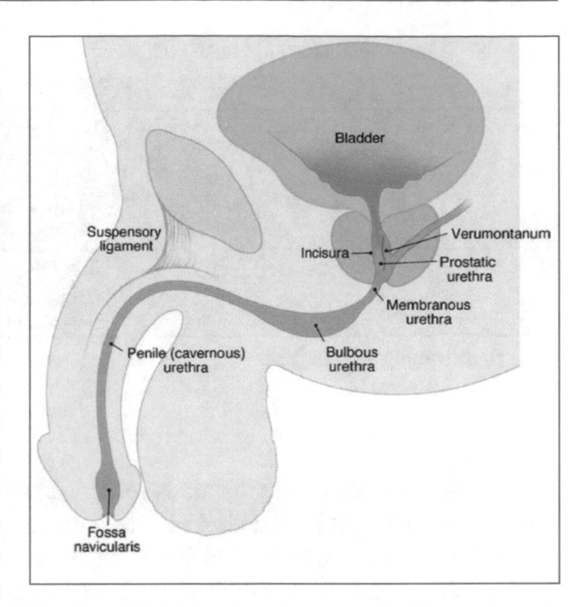

Fig. 2.4.2 Bladder and urethra anatomical image in male pelvis. Reproduced with permission from [44]

The bladder wall is composed by a tunica mucosa, tunica muscularis, tunica adventitia, and serous layer, which are involved in both the physiology and physiopathology of the organ [8]. The bladder and lower urinary tract function is that of constructing a temporary reservoir for urine. Urination, due to bladder expansion, occurs in three stages: filling stage, preurinary stage, and emptying stage. During each stage, nerve structures, detrusor muscle, sphincter systems, urethral muscles, and the pelvic plane are differentially involved. The integrity of this mechanism and related structures is necessary to maintain urinary continence. The mechanisms of urethral continence are not involved in urination itself but generate sufficient urethral resistance to prevent urine leaking through the urethra except during urination [9].

In the physiopathology of bladder damage, the bladder must be considered a serial organ, since it does not show traceable functional units. However, it is possible that each bladder layer shows its own radiosensitivity [10, 11]. Subepithelial microvascular alterations with protein migration and massive transforming growth factor beta production represent the early changes induced by ionizing radiation; subsequent accumulation of type I and III collagen fibers in the bladder wall occur [12].

Acute effects of irradiation on the bladder consist of exceeding epithelium exfoliation, leading to urothelium rupture followed by bladder vulnerability to infections and traumas. Acute symptoms are day and night pollakiuria, dysuria, cystalgia, urgent urination, and hematuria. These symptoms occur in up to 40% of patients and normally disappear 6 weeks after the end of radiotherapy course. In the chronic phase, late effects seem to be caused by degeneration of the detrusor muscle and collagen-type fibrosis across muscular layers. Obliterating endoarteritis asso-

Table 2.4.2 Physiopathology of bladder damage

Anatomical structure	Damage mechanism
Urothelium	Urothelium exfoliation → epithelial rupture (acute) Microvascular epithelial damage → protein loss, TGF-beta production → edema and inflammation (acute) Obliterating endoarteritis → wall ischemia → hematuria and or fistula (late)
Obturator muscle	Smooth-fiber degeneration deposit of collagen fibers types I and III among muscular layers } Reduced compliance

TGF transforming growth factor

ciated with bladder-wall ischemia is responsible for hematuria and/or fistula; clinical consequence is the loss of the bladder's functional capacity. Late effects usually occur within 2 years after the end of radiotherapy but sometimes occur even later – even 10 years after treatment.

Frequency and severity of acute and late effects on the bladder are a function of both treatment site and dose [12, 13] (Table 2.4.2).

2.4.3 Corpora Cavernosa, Corpus Spongiosum, and Penile Bulb

The two corpora cavernosa and the corpus spongiosum (corpus cavernosum urethrae) are the three erectile structures forming the penis. Both corpora cavernosa, equal and symmetric, show an approximately cylinder-like shape with thin terminations. In correspondence to the radix of the penis, they are separated from each other; however, they are strongly connected in the penis corpus. Each corpora cavernosa originates through a sharpened root in correspondence with the related ischiopubic branch, just in front of the ischial tuberosity. It continues in the anterior–superior direction inside the penis lodge, adjacent to the inner part of the ischiopubic branch itself, adjacent to the periosteum, and wrapped as if in a sheath by the ischiocavernous muscle on the same side. The corpora cavernosa converge at the level of the subpubic arch, where their medial surfaces come into contact, with interposition of connective fascia (penile septum). The corpora cavernosa form a longitudinal groove both dorsally and ventrally, along the entire length of the penis corpus; the dorsal groove opens into the dorsal vein of the penis, whereas the ventral groove opens into the corpus spongiosum. Toward the distal terminations, the corpora cavernosa become thinner and terminate with a blunt apex encapsulated by the glans penis. The corpora cavernosa are composed of a fibrous coating (tunica albuginea) and cavernous, or erectile, tissue.

The corpus spongiosum, unpaired and median, runs for almost its total length beside the spongious portion of the urethra, forming around it a sort of sleeve. It originates between the root of the corpora cavernosa with a swelling, urethral bulb, adhering intimately to the inferior surface of the urogenital diaphragm covered by cavernous bulb muscles. The posterior termination of the bulb is about 1-1.5-cm from the anal canal and shows a more or less marked vertical groove in the middle. The superior part of the bulb is obliquely crossed, showing a dilation called the "sulcus of the bulb" [14] (Fig. 2.4.3).

In erection physiology, three anatomical structures are mainly involved: inner pudendal artery, corpora cavernosa, and neurovascular bundle. The inner pudendal artery originates from the inner iliac artery, providing the three main arteries of the penis passing through the urogenital diaphragm: urethral bulb artery, cavernous or deep artery, and dorsal artery. The corpora cavernosa comprise the expandable erection tissue. Its base is surrounded by the ischiocavernous muscle. Relaxation of this muscle, in combination with inner pudendal artery dilation, results in rapid filling of the corpora cavernosa. Due to

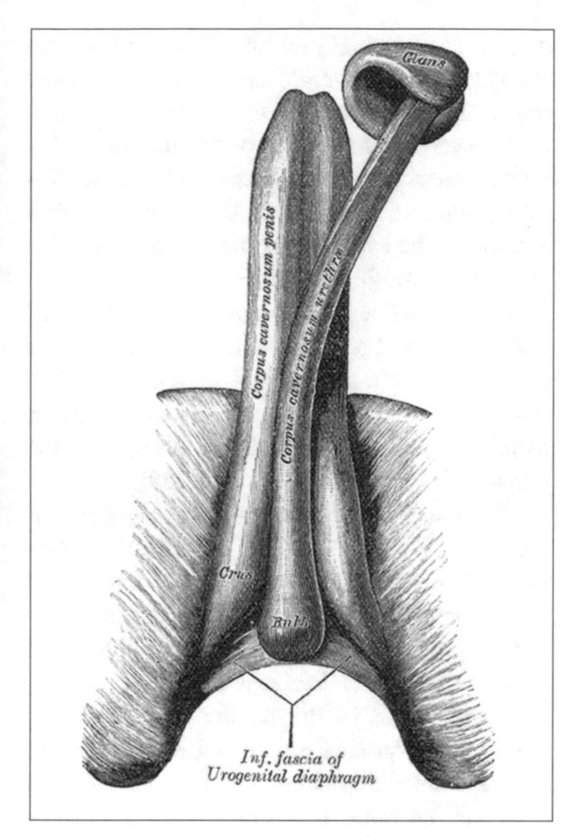

Fig. 2.4.3 Anatomical configuration of the penis (Gray's Anatomy, 1918). Reproduced with permission from [46]

the venous anatomy, the expanded corpora cavernosa compress the veins, restricting their outflow. The neurovascular bundle originates between the prostate and rectum, giving origin to two cavernous nerves. The inferior cavernous nerve passes through the genitourinary diaphragm and together with the dorsal vein of the penis crosses the diaphragm opening just behind the symphysis [15–18].

Radiotherapy to the pelvis can cause erectile dysfunction (ED) or impotence. ED incidence reported in the literature ranges from 20% to 90% and depends on several factors: technique and radiotherapy dose, instruments used to evaluate dysfunction, and length of follow-up. Direct and indirect demonstrations in the literature show a vascular mechanism at the origin of postactinic impotence, mainly related to artery damage [19–22]. The exact mechanism of ED remains unknown due to the complex interaction of vascular and neuronal factors [23–25]. Also, the penile bulb, though not an essential structure in obtaining and maintaining erection, can be involved in ED and can be a surrogate of the dose received at that level [26–28] (Table 2.4.3).

2.4.4 Urogenital Diaphragm

The urogenital diaphragm (UD), or urogenital triangle, is one of the fascia–muscular formations composing the perineum. It is located inferiorly to the pelvic diaphragm and is represented by a triangular-shaped, muscular, aponeurotic layer 1-cm thick and enclosed between the two ischiopubic branches so as to close the anterior portion of the inferior narrow side of the small pelvis. The pelvic diaphragm shows a median incisure delimited by median margins of pubococcygeus muscles (Fig. 2.4.4).

In males, the urogenital triangle is crossed by the urethra (membrane portion) and contains Cowper bulbourethral glands; in females, it is crossed by urethra and vaginal opening partially to Bartholin's vestibular glands.

The UD is composed of deep transverse perineal muscle and striated urethral sphincter mus-

Table 2.4.3 Physiopathology of erectile damage

Anatomical structure	Damage mechanism	
Penile bulb	Vascular damage → reduced blood flow to erectile tissue (vein occlusion) → reduced dilation of corpus spongiosum → reduced penile rigidity	
Neurovascular bundle	Vascular damage → reduced flow in pudendal inner artery and pudendal subsidiary arteries Nerve damage → reduced nitric oxide level	Reduced dilation of corpora cavernosa

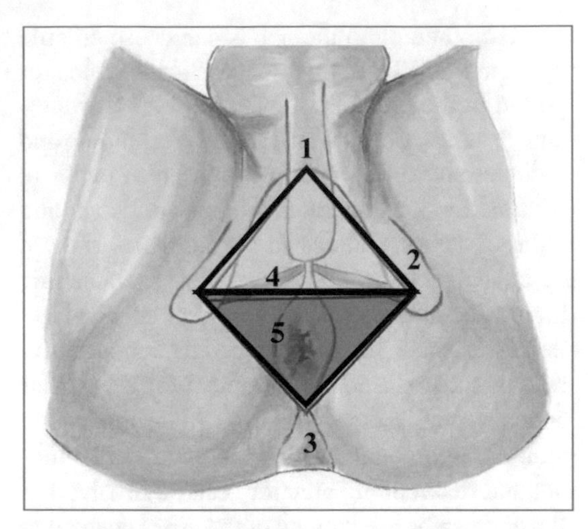

Fig. 2.4.4 Anatomical configuration of urogenital diaphragm. *1* Pubic symphysis, *2* ischial tuberosity, *3* coccyx, *4* urogenital triangle, *5* anal tiangle. Drawing by Danila Trignani

cle, which are covered both superiorly and inferiorly by an aponeurotic fascia called the perineal fascia media.

The deep transverse perineal muscle, pair muscle, arises from the branch of the ischium. Medially, this muscle becomes tendinous and crosses the one on the opposite side forming the central tendinous point of the perineum.

The striated sphincter muscle of the urethra surrounds the initial portion of the urethra, also partially covering the prostate apex in males and the vagina in females. The urogenital triangle fascias are very strong. Laterally, the inferior lamina is fixed on ischiopubic branches, whereas the superior lamina continues with the obturator fascia. Anteriorly the two laminas meet under the pubic symphysis, forming the transverse ligament of the preurethral perineum, separated from the curved pubis ligament by a thin incisure crossed by the deep dorsal vein of the penis and clitoris (Fig. 2.4.5a,b). Posteriorly, both fascias join together, forming – with other aponeurotic structures – the perineal body, or tendinous center [29].

The musles of the urogenital triangle are innervated by pudendal nerve branches; contraction of the striate sphincter muscle of the urethra determines the forced closing of the urethra itself, whereas that of the deep transverse perineal muscle determines the pelvic-floor resistance, creating tension in the tendinous center [29]. Pelvic-floor muscles are therefore involved in normal fecal and urination continence as well as in sexual functions. It was observed – for example, following external beam radiotherapy on the prostate – that such muscles are not equally exposed to radiation doses.

Although these organs all indistinctively seem to be involved in the development of late anorectal toxicity, incontinence appears to correlate to

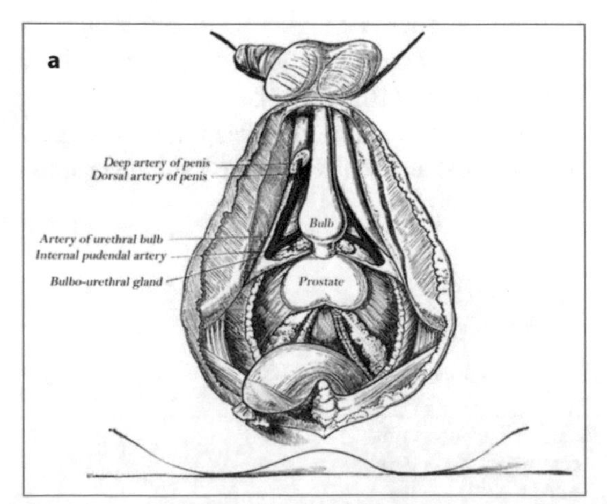

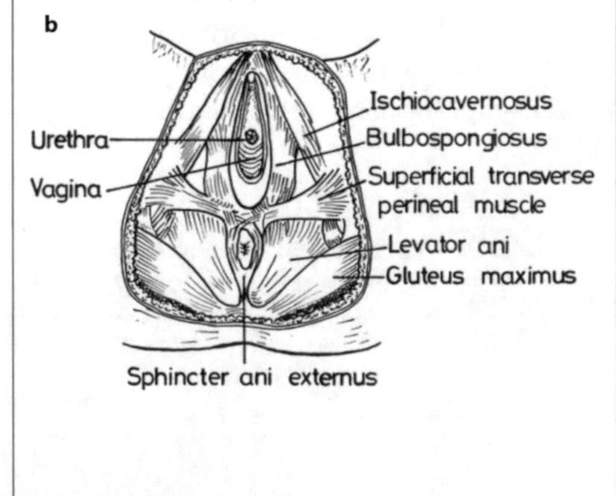

Fig. 2.4.5a, b Anatomical configuration of the perineum. **a** Male; **b** Female. Reproduced with permission from [46]

Table 2.4.4 Physiopathology of urogenital diaphragm damage

Anatomical structure	Damage mechanism
Muscle tunicae	Reduced elastic fibers → collagen fiber deposit → rigidity → reduced muscle compliance (fecal and urinary incontinence)

specific muscular structures of the pelvic floor: the puborectalis muscle, the levator ani muscle, and the internal and external anal sphincters. It has been suggested that the puborectalis muscles are the most directly responsible for fecal incontinence [30–34] (Table 2.4.4).

2.4.5 Femoral Head

The femoral head represents the proximal termination of the femur, forming by itself the thigh skeleton. It has a round shape and is set in the acetabulum, with which it articulates. In its center is a depression, the fovea capitis femoris, where the round ligament of the femur inserts connecting it to the bottom of the acetabulum; it is supported by a bone segment called the anatomical neck, at the base of which the two femoral trochanters are located (Fig. 2.4.6).

The femoral head is typically composed of spongious tissue formed by thin trabeculae, or spicules, which delimit intercommunicating (medullary) cavities filled with hemopoietic marrow. Trabeculae consist of osteons, tiny tubular structures that contain the medullary cavities. These cavities are surrounded peripherally by unorganized lamellae, or concentric bone layers, called the endosteum [35].

The head of the femur is an OAR from radiotherapy to pelvic tumors. Ionizing radiation causes both direct and indirect effects to bone. These effects are associated with vascular changes [36]. The etiopathogenetic mechanism is described and reported in paragraph 2.2.1 (Table 2.4.5).

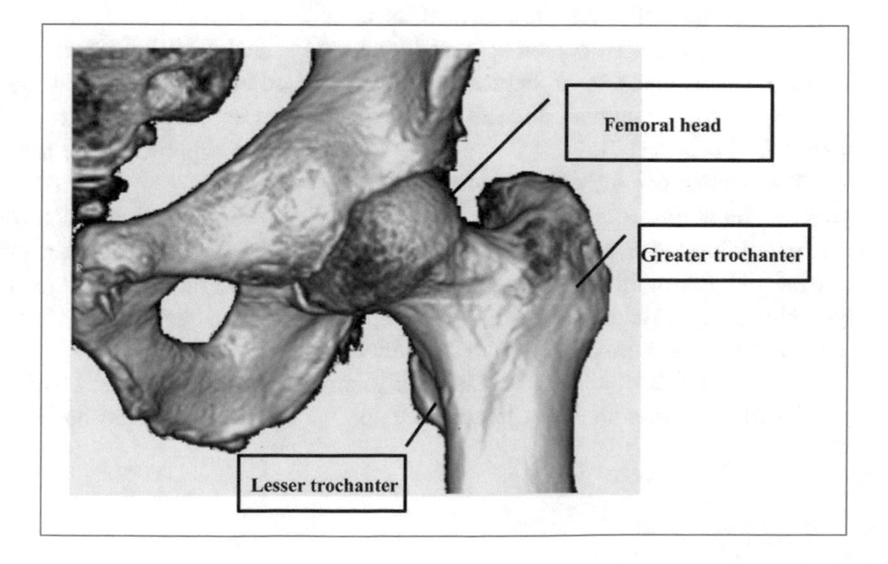

Fig. 2.4.6 Anatomical reconstructed image of the femoral head and its articulation

Table 2.4.5 Physiopathology of femoral-head bone damage

Anatomical structure	Damage mechanism
Osteoblasts	Reduced number of osteoblasts → reduced collagen fibers Alkaline phosphatase → osteopenia
Microcirculation	Vascular occlusion → necrosis/osteitis/fracture

Table 2.4.6 Physiopathology of ovarian damage

Anatomical structure	Damage mechanism	
Granulosa cells	DNA damage → cell death ↓ Hormone depletion Follicle number reduction Residual follicle atresia	 Menopause Ovary insufficiency Infertility

2.4.6 Ovaries

Ovaries are the female gonads and carry out a double function: the gametogenic function, which produces germinal cells, or oocytes; and the endocrine function, which secretes steroid hormones. Ovaries are paired organs, each being located on either side of the uterus close to the lateral wall of the small pelvis. Their shape and size are similar to a large almond, with major vertical axis positioned on a sagittal plane. The ovary is not covered by visceral peritoneum but by a particular epithelium, called the germinal epithelium.

The ovary has a variable position, as it can follow uterine shifts. Nevertheless, in its usual position, the ovary is located so that its lateral surface corresponds to a depression (ovarian groove) in the posterolateral wall of the small pelvis [37] (Fig. 2.4.7).

The ovaries begin their endocrine and gametogenic functions in puberty. These activities, characterizing the female physical maturity period, have a cycle of 28 days depending on pituitary gonadotropins. Abdominal, pelvic, and spinal-cord irradiation is associated with an increased risk of ovarian insufficiency and infertility, primarily when ovaries are in the treatment field.

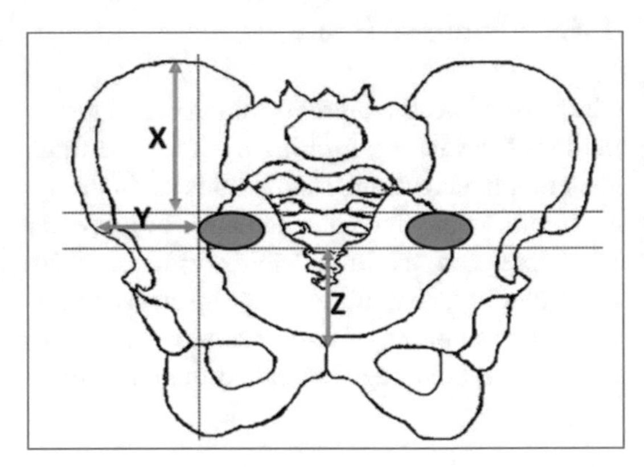

Fig. 2.4.7 Graphic representation of ovarian position in relation to pelvic osseous landmarks. Reproduced with permission from [47]

The initial target of irradiation damage is probably represented by granulosa cells that cover ovarian follicles while growing; these follicles support oocyte function during maturation [38, 39].

Ovary transposition (ovariopexy) outside the field of treatment can preserve ovary function and might be considered in females under reproductive age [40–43] (Table 2.4.6).

Acknowledgement This chapter has been written with the contribution of Marianna Trignani.

References

1. Balboni GC, Bastianini A, Brizzi E et al (1991) Anatomia Umana. Third Edition. Edi Ermes, Vol. 2, pp 156–163
2. Brading AF, Ramalingam T for the Oxford Continence Group. (2006) Mechanisms controlling normal defecation and the potential effects of spinal cord injury. 152:345–358.
3. Sedgwick DM, Howard GC, Ferguson A (1994) Pathogenesis of acute radiation injury to the rectum. A prospective study in patients. Int J Colorectal Dis 9:23–30
4. Yeoh EE, Holloway RH, Fraser RJ et al (2004) Anorectal dysfunction increases with time following radiation therapy for carcinoma of the prostate. Am J Gastroenterol 99:361–369
5. Yeoh EK, Holloway RH, Fraser RJ et al (2009) Anorectal function after three- versus two-dimensional radiation therapy for carcinoma of the prostate. Int J Radiat Oncol Biol Phys 73:46–52
6. Varma JS, Smith AN, Busuttil A (1986) Function of the anal sphincters after chronic radiation injury. Gut 27:528–533
7. Maeda Y, Høyer M, Lundby L, Norton C (2011) Faecal incontinence following radiotherapy for prostate cancer: a systematic review. Radiother Oncol 98(2):145–153
8. Balboni GC, Bastianini A, Brizzi E et al (1991) Anatomia Umana. Third Edition. Edi Ermes, Vol. 2, pp 397–408
9. McLaughlin PW, Troyer S, Berri S et al (2005) Functional anatomy of the prostate: implications for treatment planning. Int J Radiat Oncol Biol Phys 63(2):479–491
10. Cheung MR, Tucker SL, Dong L et al (2007) Investigation of bladder dose and volume factors influencing late urinary toxicity after external beam radiotherapy for prostate cancer Int J Radiat Oncol Biol Phys 67(4)1059–1065
11. Harsolia A, Vargas C, Yan D (2007) Predictors for chronic urinary toxicity after the treatment of prostate cancer with adaptive three-dimensional conformal radiotherapy: dose–volume analysis of phase II dose escalation study. Int J Radiat Oncol Biol Phys 69(4):1100–1109
12. Pointreau Y, Atean I, Durdux C (2010) Normal tissue tolerance to external beam radiation therapy: bladder. Cancer Radiothér 14(4–5):363–368
13. de Crevoisier R, Fiorino C, Dubray B (2010) Dosimetric factors predictive of late toxicity in prostate cancer radiotherapy. Cancer Radiothér 14:460–468
14. Balboni GC, Bastianini A, Brizzi E et al (1991) Anatomia Umana. Third Edition. Edi Ermes, Vol. 2, pp 460–469
15. van der Wielena GJ, Mulhallb JP, Incrocci L (2007) Erectile dysfunction after radiotherapy for prostate cancer and radiation dose to the penile structures: a critical review Radiother Oncol 84:107–113
16. Mc. Laughlin PW, Troyer S, Berri S et al (2005) Functional anatomy of the prostate: implication for treatment planning. Int J Radiat Oncol Biol Phys 63:479–491.
17. Burnett AL, Tillman SL, Chank TSK et al (1993) Immunohistochemical localisation of nitric oxide synthetase in the autonomic innervation of human penis. J Urol 150:73–76
18. Seftel AD, Resnick MI, Boswell MV (1994) Dorsal nerve block for management of intraoperative penile erection. I Urol 151:394–395
19. Incrocci L (2005) Radiation therapy for prostate cancer and erectile (dys)function: the role of imaging . Acta Oncologica 44:673–678
20. Goldstein I, Feldman MI, Deckers PJ et al (1984) Radiation-associated impotence. JAMA 251:903–910.
21. Mittal B (1985) A study of penile circulation before and after radiation in patients with prostate cancer and its effect on impotence. Int J Radiat Oncol Biol Phys 11:1121–1125
22. Zelefsky MJ, Eid JF (1998) Elucidating the etiology of erectile dysfunction after definitive therapy for prostatic cancer. Int J Radiat Oncol Biol Phys 40:129–133
23. Mc Laughlin PW, Narayana V, Meirovitz A et al (2004) Vessel-sparing prostate radiotherapy: dose limitation to critical erectile vascular structure (internal pudenda artery and corpus cavernosum) defined by MRI. Int J Radiat Oncol Biol Phys 61:20–31
24. Lepor H, Gregerman M, Crosby R et al (1985) Precise localization of the autonomic nerves from the pelvic plexus to the corpora cavernosa: a detailed anatomical study of the adult male pelvis. J Urol 133:207–212
25. van der Wielena GJ, Mulhallb JP, Incrocci L (2007) Erectile dysfunction after radiotherapy for prostate cancer and radiation dose to the penile structures: A critical review. Radiother Oncol 84:107–113
26. Dean RC, Lue TF (2005) Physiology of penile erection and pathophysiology of erectile dysfunction. Urol Clin North Am 32:379–395
27. Roach M 3rd, Nam J, Gagliardi G et al (2010) Radiation dose-volume effects and the penile bulb. 76(3 Suppl):S130–S134
28. van der Wielen GJ, van Putten WLJ, Incrocci L (2007) Sexual function after three-dimensional conformal radiotherapy for prostate cancer: results from a dose-escalation trial. Int J Radiat Oncol Biol Phys 68:479–484
29. Balboni GC, Bastianini A, Brizzi E et al (1991) Anatomia Umana. Third Edition. Edi Ermes, Vol. 2, pp 544–551
30. Smeenk RJ, Hoffmann AL, Hopman WP et al (2011) Dose-effect relationships for individual pelvic floor muscles and anorectal complaints after prostate radiotherapy. Int J Radiat Oncol Biol Phys 83(2):636–644
31. Cooper ZR, Rose S (2000) Fecal incontinence: a clinical approach. Mt Sinai J Med 67:96–105
32. Fernández-Fraga X, Azpiroz F, Malagelada JR (2002) Significance of pelvic floor muscles in anal incontinence. Gastroenterology 123:1441–1450

33. Hazewinkel MH, Sprangers MA, van der Velden J Et al (2010) Long-term cervical cancer survivors suffer from pelvic floor symptoms: a cross-sectional matched cohort study. Gynecol Oncol 117:281–286
34. Plotti F, Calcagno M, Sansone M et al (2010) Long-term cervical cancer survivors suffer from pelvic floor symptoms. Gynecol Oncol 119(2):399; author reply 399–400
35. Balboni GC, Bastianini A, Brizzi E et al (1991) Anatomia Umana. Third Edition. Edi Ermes, Vol. 1, pp 279–284
36. Howland WJ, Loeffler RK, Starchman DE, Johnson RG (1975) Postirradiation atrophic changes of bone and related complications. Radiology 117:677–685
37. Balboni GC, Bastianini A, Brizzi E et al (1991) Anatomia Umana. Third Edition. Edi Ermes, Vol. 2, pp 478–501
38. Lushbaugh CC, Casarett GW (1976) The effects of gonadal irradiation in clinical radiation therapy: a review. Cancer 37:1111–1125
39. Hamre MR, Robison LL, Nesbit ME et al (1987) Effects of radiation on ovarian function in long-term survivors of childhood acute lymphoblastic leukemia: a report from the Childrens Cancer Study Group. J Clin Oncol 5:1759–1765
40. Clayton PE, Shalet SM, Price DA et al (1989) Ovarian function following chemotherapy for childhood brain tumours. Med Pediatr Oncol 17:92–96
41. Grigsby PW, Russell A, Bruner D et al (1995) Late injury of cancer therapy on the female reproductive tract. Int J Radiat Oncol Biol Phys 31:1281–1299
42. Meirow D, Nugent D (2001) The effects of radiotherapy and chemotherapy on female reproduction. Hum Reprod Update 7(6):535–543.
43. Lo Presti A, Ruvolo G, Gancitano RA, Cittadini E (2004) Ovarian function following radiation and chemotherapy for cancer. Eur J Obstet Gynecol Reprod Biol 113(Suppl 1):S33–40
44. Levin TL, Han B, Little BP (2007) Congenital anomalies of the male urethra. Pediatric Radiology 37:851-862
45. Hasler W, Boland CR, Feldman M (2002) Gastroenterology and Hepatology. Volume 2. Current Medicine US
46. Rosenblum JL, Burnett AL (2013) Microsurgical Penile Revascularization, Replantation, and Reconstruction In: Sandlow JI (ed) Microsurgery for Fertility Specialists. Springer, pp 179-221
47. Bardo Dianna ME, Black M, Schenk K, Zaritzky M (2009) Location of the ovaries in girls from newborn to 18 years of age: reconsidering ovarian shielding. Pediatric Radiology 39:253-259

Part II

Modeling and Organ Delineation in Radiation Therapy

Chapter 3
Radiation Dose Constraints for Organs at Risk: Modeling and Importance of Organ Delineation in Radiation Therapy

3.1 Introduction

The optimal goal of radiation therapy (RT) is to eliminate a maximum number of clonogenic (stem) cancer cells while producing as little treatment morbidity as possible. Often, the limit of radiation dose administered to the tumor is dictated by the tolerance to radiation of adjacent normal tissues, or organs at risk (OAR). With the advent of image-based or image-guided RT – more precisely, 3D treatment algorithms and developments in technology, including intensity-modulated (IMRT), image guided (IGRT), and adaptive RT (ART) – it has become feasible to implement dose escalation in several tumor sites. At the same time, quantitative assessment of normal tissue effects has gained more clinical relevance. Normal tissue complications occur in the context of a patient's expected longevity, and adequate follow-up after treatment is necessary to fully assess the impact of RT on OARs.

These advances have increased the need for more accurate delineation of tumor target volumes and OARs, with a quantitative evaluation of tolerance to doses of irradiation delivered to specific volumes of normal tissue, as depicted on a dose-volume histogram (DVH). In the treatment planning process, it is critical to delineate sensitive structures and give them a high priority for avoidance of radiation dose (when technically feasible) to decrease undesirable sequelae. However, adequate target coverage with a margin [planning target volume (PTV)] should not be compromised to reduce injury to OARs, except in critical situations. Further, complication-risk assessment should not be based on a single DVH constraint, and robust multimetric analysis of actual dose distribution plots will yield more precise information. This valuable imaging information will only contribute to dose–volume record activities if effective, automated segmentation methods are developed to address the laborious nature of manual segmentation. The development of model-based approaches to segmentation is congruent with an important trend toward the adoption of structured descriptions or atlases of patient anatomy [1]. Throughout the Quantitative Analysis of Normal Tissue Radiation Effects in the Clinic (QUANTEC) monograph mentioned below [2], various authors emphasize the difficulty and importance of adequate organ volume definition to arrive at more meaningful and credible radiation dose–volume parameters and OAR radiation tolerance constraints.

It is understood that normal tissue has a serial or parallel functional structure, which influence the impact of radiation effects on sequelae. In parallel architecture, the subvolumes of the organ function relatively independently, and complications develop when critical volumes are injured. In contrast, in serial architecture organs, complications occur even when small segments of the organ are damaged [2]. As emphasized by Marks et al. [2], special care should be taken when applying models, especially when clinical

The original version of this chapter was revised: The word "normal standard dose" was incorrectly captured in the second paragraph of 3.2 Synopsis of Historical Perspective, p. 50 of this book. The word should read as "nominal standard dose". This has been corrected. A correction of this book can be found at DOI 10.1007/978-88-470-5256-7_13

G. Ausili Cefaro, D. Genovesi, C.A. Perez, *Delineating Organs at Risk in Radiation Therapy*,
DOI: 10.1007/978-88-470-5257-4_3, © Springer-Verlag Italia 2013

dose–volume parameters are beyond the range of data used to generate the model. Typically, the models that have been designed are based on DVHs, which are not ideal representations of the 3D radiation doses. DVHs do not represent all organ-specific spatial radiation dose information (and hence assume all regions are of equal functional importance), and they often do not consider fractionation schemas. They are usually based on a single planning computed tomography (CT) scan, which does not account for anatomical and functional variations during therapy, interinstitutional/interobserver differences in image segmentation, dose calculation, patient populations, and the like. Preferred-beam arrangements may limit model exportability and comparability. Before introducing a predictive model into a clinical practice, it is prudent to assess whether its predictions are in accord with the clinic's treatment plans and experience.

Radiation effects are dose, fractionation, and volume dependent. In the QUANTEC reviews, fraction impact is adjusted based on the linear quadratic equation (LQE). This has inherent limitations, particularly with less than five fractions, in which the validity of the LQE has been questioned [2].

Combination RT with cytotoxic agents or targeted therapies (i.e., cetuximab) enhance radiation effects, and data quantifying that enhancement is often lacking. Type of agent, dosage, and schedules need to be taken into account [2].

3.2 Synopsis of Historical Perspective

Biological effects of radiation in matter are direct (energy deposition in biologically important molecules such as DNA) or indirect (production of a positive ion, H_2O^+, and a free electron, which undergo a series of reactions to produce free radicals, which interact with DNA molecules). These interactions lead to single or double DNA breaks and damaged bases. The resulting cellular injuries cause lethal, sublethal, or potentially lethal damage, with cell death, chromosomal abnormalities, and mutagenesis [3]. Response of normal tissue cells to ionizing radiation depends on a complex interplay of the inherent sensitivity of cells to irradiation, cellular kinetics, vascular and microenvironmental factors, and tissue organization. As noted, the serial or parallel architecture of OAR determine the type of injury and tolerance to irradiation.

Based on empirical observations, pioneers in RT such as Coutard , Regaud, Paterson [4–6], and others designed fractionated courses of irradiation to decrease normal tissue injury. Strandqvist [7] introduced the concept of dose–time relationship and the probability of tumor control in skin cancer. Ellis [8, 9] developed the nominal standard dose and Cohen and Kerrick [10] the tumor dose factor to express biological equivalency of various radiation dose-fractionation sched-ules on normal tissues and tumors. Tables were generated with tolerance doses of irradiation for specific volumes of multiple normal tissues. The LQE concept has been used to enhance our understanding of various radiation fractionation and dose schedules in the treatment of a variety of tumors and normal-tissue radiation effects [11, 12]. Application of the LQE to hypofrac-tionated stereotactic irradiation is discussed in detail by Fowler et al. [13] and Liu et al. [14]. Niemierko [15] formulated the equivalent uni-form dose (EUD), a mathematical concept that assumes that two radiation-dose distributions are biologically equivalent if they cause the same ra-diobiological effect. EUD has been used as a tool to compare treatment plans or mathematically as an optimization parameter in inverse planning systems.

Emami et al. [16] compiled data from the literature and attempted to quantify tolerance to radiation dose and defined volumes (one third, two thirds, whole organ) of a variety of normal structures. Probability of tumor control and normal tissue complication probability (NTCP) have been widely used in the literature to express these principles. Burman et al. [17] fitted a Lyman model [18] to the Emami consensus dose–volume data, thereby facilitating the use of Emami's constraints for an arbitrary fraction of a whole organ uniformly irradiated. Further, Kutcher et al. [19] proposed the so-called DVH reduction algorithm for reducing an arbitrary, nonuniform dose distribution into a partial volume receiving the maximum dose, effectively allowing the extrapolation of Emami's constraints to any dose distribution.

Trotti et al. [20] published a review of a number of adverse effects (AE) of treatment terminology grading systems, leading to the third version of the US National Cancer Institute *Common Toxicity Criteria* (CTC). It was renamed *Common Terminology Criteria for Adverse Events* version 3.0 (CTCAE v3.0), with the purpose of moving away from the term toxicity, which implies causation and does not fit the lexicon commonly used across all treatment modalities. The CTCAE v3.0 contains criteria for approximately 570 AEs, about 35 for specifying anatomical sites (e.g., fistula–rectal, esophageal, tracheal, and so on) or other subclassifications, resulting in about 900 site-specific AE criteria for grading acute and late effects. Marks et al. [2] were editors of a special issue of the *International Journal of Radiation Oncology Biology and Physics* devoted to an update of recent literature data focused on QUANTEC, incorporating more recent quantitative 3D dose–volume parameters on the subject.

On the modeling side, there is a need for improved data analysis methods and a more critical appraisal of the various dimensions of model validity [21]. A typical NTCP model fit to clinical data sets yields a relatively low statistical power of goodness of fit statistics. The log-likelihood may also be used for comparing the fit of competing models to a data set; studies have shown that competing models tend to produce very similar log-likelihood values for a given data set. For nested models (i.e., models that differ by the inclusion of one additional parameter), the difference in log-likelihood forms the basis for the likelihood ratio test, a robust test for the statistical significance of adding this parameter [21].

External validity addresses how well the model explains the variability in response seen in an independent data set, preferably from another institution. Multivariate NTCP models are often overfitted in the sense that they include too many parameters relative to the number of events analyzed. This may result in strongly correlated parameter estimates, and although such a model may pass the test for internal validity with flying colors, it often has poor external validity. Differences between institutions in the scoring of reactions, in patient demographics, in the burden of comorbidities, and in treatment characteristics may all contribute to a reduced predictive power of a model when tested in an independent data set [23]. Relatively little research has been performed on the external validity of NTCP models [37]. The fact that dose–volume constraints or NTCP models are used in clinical practice does not in itself prove that they improve cancer care from an evidence-based medicine perspective. Ultimately, the clinical utility of NTCP modeling should be tested in randomized controlled trials [21]. As emphasized by Deasy et al. and Andrew et al. [22], there is a critical need for the radiation oncology community to agree on common terminology, methodology, definition of model parameters, and statistical testing and to collaborate to create pools of data that will be available for analysis by various investigators. It is only by making published data sets available for ongoing combined analyses that we can produce powerful and validated models of quantitative normal tissue effects in the clinic [22].

3.3 Clinical Application of the Linear Quadratic Equation

The following section is extracted from a book chapter by Halperin et al. [24]. Formulations based on dose-survival models have been proposed to evaluate the biological equivalence of various doses and fractionation schedules. These assumptions are based on an LQ survival curve represented by the equation

$$\mathrm{Log_e S} = \alpha D + \beta D2$$

in which α represents the linear (i.e., first-order, nonreparable, dose-dependent) component of cell killing, and β represents the quadratic (i.e., second-order, more reparable, dose-dependent)

Table 3.1 Alpha/beta values

Early-responding tissues	α/β (Gy)	Late-responding tissues	α/β (Gy)
Skin (desquamation)	9.4–21	Brain (LD50)	2.1
Melanocytes (depigmentation)	6.5	Spinal cord	2
Lip mucosa (desquamation)	7.9	Lung (pneumonitis)	1.6–4.5
Tongue mucosa (ulceration)	11.6	Lung (fibrosis)	2.3
Jejunal mucosa (clones)	7–13	Heart (failure)	3.7
Colonic mucosa	7–8.5	Liver (clones)	2.5
Spleen	8.9	Breast (fibrosis)	4–5
Bone marrow	9	Bowel (stricture, perforation)	3.5–5
Tumors		Rectum	5
Breast cancer	3.5–4.6	Bladder (fibrosis, contraction)	5.8
Epithelial (head and neck)	8–10	Bone-cartilage	2–4
Prostate	1.5–3	Eye (cataracts)	1.2

Values represent synthesis of many publications. Modified from McBride and Withers [25]. *LD50*, median lethal dose (lethal dose, 50%)

component of cell killing. The dose at which the two components of cell killing are equal constitutes the α/β ratio. Examples of alpha/beta ratios are shown in Table 3.1 [25].

In general, tumors and acutely reacting tissues have a high α/β ratio (8–15 Gy), whereas tissues involved in late effects have a low α/β ratio (1–5 Gy). The values for α and β can be obtained from graphs in which the reciprocal of the total dose (TD) (Gy−1) and the dose per fraction (Gy) are plotted. A straight line is obtained. The interception of this line with the zero dose-per-fraction axis is proportional to α and equal to α/lnS, where S is the natural logarithm of survival. The slope is proportional to β and equal to β/lnS. The algebraic functions to derive the straight line from the reciprocal TD-per-fraction plot are provided as follows: Tumor cell survival following n fractions, each of dose d:

$$-\ln S = n\,(\alpha d + \beta d)^2$$
$$= \alpha nd + \beta nd^2$$
$$= nd\,(\alpha + \beta d)$$

Dividing both sides by TD nd:

$$-\ln S / nd = \alpha + \beta d$$

Withers et al. [26] proposed a method for using these survival-curve parameters for calculating the change in TD necessary to achieve an equal response in tissue when the dose per fraction is varied, using the α/β ratios. This calculation accounts only for the effect of repair of cellular injury. Isoeffect curves vary for different tissues. A biologically equivalent dose (BED) can be obtained using this formula:

$$BED = \ln S / \alpha.$$
$$BED = nd\,[1 + d / (\alpha/\beta)]$$

If one wishes to compare two treatment regimens, the following formula can be used:

$$Dr / Dx = (\alpha/\beta + dx) / (\alpha/\beta + dr)$$

in which *Dr* is the known TD (reference dose), *Dx* is the new TD (with different fractionation schedule), *dr* is the known fractionation (reference), and *dx* is the new fractionation schedule. Let us consider an example of the use of this formula (with some reservations). Suppose 50 Gy in 25 fractions is delivered to yield a given biologic effect. If one assumes that the subcutaneous tissue is the limiting parameter *r* (late reaction), it is desirable to know what the TD to be administered will be using 4-Gy fractions. Assume α/β for late fibrosis equals 2 Gy. Using the above formula:

$$Dx = Dr\ (\alpha/\beta + dr) / (\alpha/\beta + dx).$$

Thus:

$$Dx = 50\ Gy\ (5 + 2 / 5 + 4) = 50\ (7/9) = 39\ Gy.$$

The basic LQ equation addresses the inactivation of a homogeneous population of cells. One should be wary, however, of accepting the basic equation as being completely accurate. Because it is likely that accelerated repopulation of tumor clonogens occurs during the course of fractionated RT, and that cell-cycle redistribution and reoxygenation also occur, we should consider how these factors can be accounted for in the formula. Repopulation may be accounted for, in broad approximation, by describing the number of clonogens (N) at time t as being related to the initial number of clonogens (No). Then,

$$N = No^{e\lambda t.}$$

The parameter λ determines the speed of cell repopulation and is given by

$$\lambda = loge^2 / Tpot = 0.693 / Tpot$$

where *Tpot* is the effective doubling time of cells in the tumor. If we ignore spontaneous cell loss, then Tpot is approximately the same as the measurable *in vitro* doubling time of tumor cells. Reported values of Tpot are 2–25 days, with a median value of approximately 5 days. For late-responding tissues, Tpot is so large that λ is effectively zero. Incorporating the allowance for tumor proliferation, with t representing time, the LQ equation becomes:

$$E = nd\ (1 + d / \alpha/\beta) = 0.693t / \alpha Tpot$$

Now let us assume α/β of 10 for an acutely reacting tissue, such as a tumor, and an α of 0.3 with a Tpot of 5. The BED of 70 Gy with 2-Gy/fraction, five fractions per week, in 46 days, is

$$BED = 70\ (1 + 0.2) = 84\ Gy_{10}.$$

Now let us add the correction for tumor repopulation during the course of treatment:

$$E/\alpha = 70\ (1 + 2/1) - (0.693/0.3) \times 46 / 5$$
$$BED = 84 - 21 = 63 Gy_{10}.$$

The decrease in clonogens by RT is attenuated, in part, by repopulation of the surviving clonogens. In the LQ equation, redistribution in the cell cycle and reoxygenation may be modeled by a single term called resensitization. Immediately after a dose of radiation, the average radiosensitivity of the cell population falls and then gradually returns to greater sensitivity. In contrast to tumor proliferation, resensitization probably increases as overall treatment time increases. Not enough is known about the clinical importance or resensitization to make it useful to incorporate a numeric value for it in the LQ formula. The LQ model can be used to construct a biologically oriented dose-distribution algorithm for clinical RT [27].

3.4 Clinical Data on Radiation Tolerance Constraints of Organs at Risk

3.4.1 Brain

Emami et al. [16] noted that it is highly uncommon to observe radiation injury to the brain with doses of ≤50 Gy unless high doses per fraction are administered. For fractionated RT with a fraction size of <2.5 Gy, incidence of radiation necrosis of 5% and 10% is predicted to occur at a median BED+ of 120 Gy and 150 Gy, respectively, corresponding to conventionally fractionated doses of 72 and 90, respectively. Cognitive dysfunction in children is usually seen for whole-brain doses of ≥18 Gy. For twice-daily fractionation, a steep increase in toxicity appears to occur when the biologically effective dose is >80 Gy.

For large fraction sizes (≥2.5 Gy), the incidence and severity of toxicity (4–15%) is unpredictable. For single-fraction radiosurgery, a clear correlation has been demonstrated between the target size and the risk of adverse events. The volume of brain receiving ≥12 Gy has been shown to correlate with both the incidence of radiation necrosis and asymptomatic radiological changes. Toxicity increases rapidly once the volume of the brain exposed to >12 Gy is >5–10 cm^3. The Radiation Therapy Oncology Group (RTOG) conducted a dose-escalation study to define the maximal dose for targets of different sizes in patients who had previously undergone whole-brain irradiation. The maximal tolerated dose for targets 31–40 mm in diameter was 15 Gy; for targets 21–30 mm 18 Gy, and for targets <20 mm >24 Gy [28]. Functional areas of the brain (brain stem, corpus callosum) require more stringent limits. Substantial variation among reported outcomes from different centers has prevented predictions regarding toxicity–risk [29].

In patients treated for brain metastasis, fractions of 3–6 Gy result in greater cognitive and memory deficits compared with 2–2.5 fractions when 35–40 Gy are administered [30].

3.4.2 Hypothalamus, Pituitary

The hypothalamus–pituitary axis, with its complex hormonal regulatory functions, is sensitive to moderate to high doses of radiation (30–50 Gy), dependent on the age of the patient and fractionation schedules.

Hypopituitarism is reported in 20% of patients 5 years following treatment for pituitary adenomas (50 Gy in 2-Gy fractions) but it increases to 50–80% by 20 years [31]. In 312 patients treated for head and neck cancer in whom the pituitary as irradiated, at a median follow-up of 5.6 years, 44 (14%) experienced some clinical hypopituitarism; in 68 asymptomatic patients, 33.8% had subclinical dysfunction [32]. Snyers et al. [33], in a study of 76 patients with sinonasal tumors treated with surgery and irradiation, 24% had true hypothalamic–pituitary–adrenal (HPA) deficits (mean dose to hypothalamus 51.6–56 Gy), and 57% had subclinical deficits (mean dose to hypothalamus 40 Gy and to pituitary 62 Gy).

In patients treated with hypofractionated stereotactic RT (dose ranging from 20 Gy to 40 Gy, one to four fractions), the average incidence of hypopituitarism ranges from 6% to 52% by 5 years [30]. In a study of 130 patients treated with the gamma knife for pituitary adenomas (mean dose 15–25 Gy), parameter predictors for additional pituitary deficits were dose to the pituitary gland 15.7 Gy and to the stalk 7.3 Gy [34].

3.4.3 Brain Stem

Emami et al. [16], in their compilation study, noted that estimated radiation doses of 60, 53, or 50 Gy to one-, two-, or three-thirds of the brain stem, respectively, yielded a 5% probability of injury at 5 years and doses ≥65 Gy induced a 50% probability of adverse effects. More recent literature data suggests that 55 Gy with 2-Gy fractionation photon irradiation is the threshold for injury to the brainstem [35], although in definitive irradiation for nasopharyngeal carcinoma, small segments of the brain stem receive 60 Gy. With hypofractionated stereotactic irradiation, a 16-Gy single dose is an important factor for late cranial neuropathy, which is related to brain stem damage [36].

3.4.4 Optic Nerve and Optic Chiasm

Radiation-induced optic neuropathy (RION) has been observed with doses >50 Gy, depending on fractionation schemas. Parsons et al. [38] observed a risk of optic neuropathy of 8% with doses between 60 and 73 Gy with fractionation of 1.65 and 1.80 Gy, but it increased to 41% with fractions >1.95 Gy. Van den Berg et al. [39] noted, in a review of several reports of patients irradiated for acromegaly, that 75% of patients developing optic neuropathy received fractionation doses >2 Gy. Hammer et al. [40] reported that chiasm injury in 4/87 patients (4.6%) irradiated for pituitary tumors occurred at doses of 42.5 Gy with fractions of 2–2.8 Gy.

Several proton series have reported a very low incidence of RION, with a threshold of 55–60 cobalt Gray equivalent, consistent with that of photons [41]. Some studies have shown that optic nerve tolerance to hypofractionated stereotactic irradiation is 8 Gy, single fraction. Tishler et al. [42] found optic nerve injury in 24% of patients treated with >8 Gy, in contrast to no injuries with lower doses. However, other studies have not reported optic neuropathy with doses up to 10 Gy [43].

3.4.5 Spinal Cord

The cervical spine has been shown to tolerate doses of 50 Gy [44], whereas the thoracic spine tolerance has been traditionally held at 45–47 Gy, with 1.8–2 Gy fractions (BED 100 Gy^2) for volumes <10 cm [45–46]. Higher doses have been associated with cord myelopathy in 35–50% of patients.

Fractionation plays a critical role in the development of spinal cord radiation injury. With conventional fractionation of 2 Gy per day, including the full-cord cross-section, TDs of 50 Gy, 60 Gy, and 69 Gy are associated with a 0.2%, 6%, and 50% rate, respectively, of myelopathy [47].

The advent of hypofractionated stereotactic body RT (SBRT) generates new challenges in OAR tolerance to irradiation. Limited data by Sahgal et al. [46] in five patients strongly suggest that a single spinal cord dose should not exceed 10 Gy. Daly et al. [48], in a study of 17 patients with spinal hemangioblastoma treated with SBRT (median dose 20 Gy single dose), noted only one instance of myelopathy (4%). They concluded that traditional radiobiological models used to estimate cord NTCP may not apply to SBRT and that further research is needed.

3.4.6 Brachial Plexus

Brachial plexopathy has rarely been reported; when it occurs, it is usually in patients irradiated for breast cancer with doses per fraction >3 Gy. In a series of patients treated at Peter MacCallum Cancer Center (Melbourne, VIC, Australia) with 63 Gy in 12 fractions or 57.75 Gy in 11 fractions (5.25-Gy fractions) resulting in doses to the brachial plexus of 55 Gy or 51 Gy, respectively, the incidence of neurological symptoms was 73% and 15%, respectively, at a maximum follow up of 30 months [49]. In contrast, with 50 Gy in 2-Gy fractions, the incidence on brachial neuropathy is ≤1%.

Concurrent chemotherapy does increase the incidence of plexopathy. Pierce et al. [49] reported 0.4% (3/724) with radiation alone (≤50 Gy) and 3.4% (10/267) when combined with chemotherapy. When the radiation axillary dose was >50 Gy, the incidence of plexopathy was 3% (2/63) and 8% (5/63), respectively. In 95 patients with head and neck cancer irradiated to doses of 60–70 Gy, in 31 receiving concurrent chemoradiation, the incidence of plexopathy was 16% (6/38) when patients followed up for <1year were excluded [51]. Brachial plexopathy was noted in seven of 36 patients (19%) with apical lung tumors treated with SBRT (30–72 Gy in three to four fractions) [52].

3.4.7 Eye

The lenses are extremely sensitive to irradiation; single doses of 10 Gy or fractionated doses of 16–20 Gy will induce cataracts in up to 50% of patients. Henk et al. [53] reported 74% lens opacities in patients who received lens doses of 30 Gy in 20 fractions.

3.4.8 Retina

Radiation-induced retinopathy has been observed in patients irradiated for Graves' ophthalmopathy at doses <20 Gy in 2-Gy fractions [43], although no retinal injury was reported in 311 patients who received doses of 20–30 Gy with 2-Gy fractions or in 59 treated with 25–40 Gy in 2-Gy fractions. Parsons et al. [38] reported retinal damage with doses >45 Gy. There is a steep response curve, and with higher TDs or dose per fraction, the probability of injury increases to 50% in 5 years. Monroe et al. [54] noted that twice-daily hyperfractionation decreases the incidence of retinopathy; in patients receiving doses >50 Gy, the incidence was 13% versus 37% for once-daily fractionation.

3.4.9 Ear

Sensorineural hearing loss (SNHL) has been observed with doses >60 Gy to the inner ear. Bhandare et al. [55] reported clinical hearing loss in 37% of patients receiving 60.5 Gy in contrast to 3% with lower doses. Chan et al. [56] described an incidence of SNHL at a high frequency in 55% of patients with nasopharynx carcinoma treated with chemoradiation therapy versus 33.3% in a group treated with irradiation alone (p = 0.02) but not at a low frequency (7.9% vs. 16.7%, p = 0.17). With hypofractionated stereotactic schedules, hearing preservation has been reported with doses <16 Gy [57].

3.4.10 Parotid and Other Salivary Glands

Xerostomia is induced by fractionated irradiation with doses >25 Gy, with partial recovery of function within 2 years after exposure, depending on dose [93]. In a study by Blanco et al. [58], patients irradiated to both left and right parotid with mean doses >25 Gy tended to have poor salivary function. If one or both parotids received significantly <25 Gy, the salivary capacity typically exceeded the 25% level. According to Meirovitz et al. [59], for complete salivary production recovery after 24 months, the volume of contralateral parotid receiving >40 Gy should be <33%.

IMRT has enhanced the sparing of salivary-gland function when compared with conventional or 3D conformal radiotherapy (3D-CRT) techniques [60–61]. Amifostine was found to decrease salivary-gland flow when the parotid gland received <40.6 Gy with 3D-CRT [62].

When parotid or level II cervical lymph node metastasis is present, it is not a good practice to spare the adjacent salivary gland, as failures at that site have been reported [63].

The submaxillary salivary glands contribute 25–30% of the salivary flow. Wang et al. [64], in a study of 52 patients with head and neck cancer treated with IMRT, evaluated the effect on xerostomia of sparing or not a contralateral submaxillary gland (cSMG). Xerostomia grades at 2 and 6 months post-IMRT were significantly lower among patients in the cSMG-sparing group than in the cSMG-unspared group, but differences were not significant at 12 and 18 months.

3.4.11 Mandible and Temporomandibular Joint

Dental hygiene and volume of mandible irradiated, in addition to dose and fractionation, play an important role in the induction of osteonecrosis of the mandible. Cooper et al. [65] found no incidence with doses <65 Gy and 80% with doses >75 Gy, with osteonecrosis being more frequent in dentulous than edentulous patients.

Trismus at therapeutic dose levels (70 Gy) has been reported in 5–38% of patients irradiated for head and neck cancer, with varying degrees of impact on quality of life [66, 67].

3.4.12 Larynx

Damage to the larynx (chondronecrosis) has been observed with doses >70 Gy, with conventional fractionation (2 Gy); the incidence with usual therapeutic doses is <1%.

Cox et al. [68] evaluated the incidence of grade 2 edema of the larynx using RTOG toxicity criteria. For each 1% of laryngeal volume that received >50 Gy with 2-Gy fractions, there was a 3% increase in the rate of grade 2 or greater laryngeal edema. Sanguineti et al. [69] reported that the mean laryngeal dose or percentage of volume receiving ≥50 Gy and neck node stage were independent predictors of laryngeal edema. The investigators suggested that the percentage of volume receiving ≥50 Gy and the mean laryngeal dose should be kept as low as possible, ideally <27% and dose <43.5 Gy, respectively, to minimize the edema.

Bae et al. [70], in a randomized study of 127 patients with larynx and pharynx cancer – all but two of whom were treated with conventional RT 62–80 Gy (median dose 70 Gy) and 2-Gy per fraction (46% received concurrent chemotherapy), noted laryngeal edema grade 1 in 28 (21%) and grades 2–4 in 56 (44%); grades 3–4 laryngeal edema were more frequent in patients with supraglottic and subglottic tumors and more advanced stages (T2–4).

Current models suggest that a 50% probability of normal tissue complications is observed at mean doses of 60 Gy to these structures. The limitations of these models include treatment variables, the most important of which is concurrent chemotherapy, variations in tumor location, and pretherapy dysphagia [71].

3.4.13 Pharyngeal Constrictor Muscles

Practically all patients irradiated for head and neck cancers, especially when combined with concurrent chemotherapy, experience varying degrees of dysphagia and >50% exhibit aspiration. Eisbruch et al., Feng et al., and Teguh et al. [72–74] emphasized the importance of radiation doses >40 Gy to the pharyngeal constrictor muscles in the induction of this sequela. Caglar et al. [75] reported no aspiration or stricture when <21% of the larynx received doses >50 Gy or when <51% of the pharyngeal constrictors received >50 Gy. Caudell et al. [76] found that doses >60 Gy to ≥12% of the constrictors muscles were associated with increased need for percutaneous gastrostomy tubes and that a dose >65 Gy to >33% of the superior pharyngeal or >75% of the middle pharyngeal constrictor was correlated with greater incidence of pharyngoesophageal stricture requiring dilatation.

3.4.14 Thyroid Gland

The incidence of hypothyroidism varies with tumor stage, surgery type, radiation dose to the thyroid, concurrent chemotherapy administration, and preexisting thyroiditis. Hypothyroidism has been noted in about 30% of patients irradiated for head and neck tumors when the thyroid gland received doses >45 Gy, particularly when concurrent chemoradiation was administered [77–78]. Bhandare et al. [79] reported that when the dose to the gland was >45 Gy, the incidence of hypothyroidism increased from 60% to 76% at 5 years.

With IMRT, an effort should be made to decrease the volume of and radiation dose to the thyroid.

3.4.15 Heart

Radiation injury to the heart may be recognized in the pericardium, myocardium, heart valves, or coronary arteries in patients treated for Hodgkin's disease, carcinoma of the breast, or lung cancer. Perfusion abnormalities have been documented with doses in the range of 45 Gy. Increased incidence of pericarditis in Hodgkin's disease patients or myocardial infarction in women receiving RT for left-sided breast cancer, particularly when the internal mammary nodes are irradiated (doses >40Gy), have been noted [80]. For pericarditis, according to Wei et al. [81], the risk increases with a variety of dose parameters, such as mean pericardium dose >26 Gy, and V_{30} >46%. Dose per fraction is a factor in pericarditis induction, as is the volume of heart irradiated [82].

In patients treated with breast-conserving therapy, with a median time to event of 10–11 years, the incidence of cardiovascular disease (CVD) overall was 14.1%, of ischemic heart disease 7.3%, and for other types of heart disease 9.2%. The incidence of CVD was 11.6% in patients with right-sided breast cancer compared with 16.0% in left-sided cases. The maximum heart distance (MHD) did not correlate with CVD incidence, although patients with MHD >3 cm (representing larger volume of heart irradiated) were at a higher risk [83]. Concurrent administration of cytotoxic agents, particularly anthracyclines, increases cardiac radiation toxicity.

Feng et al. [84] published an atlas of cardiac anatomy volumes based on CT data to improve contour accuracy in defining volumes of heart irradiated in patients treated for breast cancer in order to decrease the dose to various structures.

3.4.16 Lung

Radiation pneumonitis (RP) and, later, pulmonary fibrosis, are common sequelae of irradiation to the thorax. Because lung volumes vary with breathing, there is ambiguity in defining lung DVH-based parameters. In most reports, dosimetric information is based on CT images obtained during free breathing. The dosimetric parameters would change had these scans been obtained at specific phases of the respiratory cycle (inspiration, expiration). Segmentation of a thoracic scan can be challenging. There is uncertainty regarding how much of the bronchus should be defined as "lung," and the lung edges may vary with the window/level setting. Thus, volume-based parameters will vary between investigators. The accuracy of any autosegmenting tool should be carefully assessed, especially to ensure that portions of atelectatic lung or tumor at soft-tissue interfaces are not inadvertently omitted from the lung. During RT planning, total lung volume is usually defined excluding the gross tumor volume (GTV). Excluding the planning target volume (PTV) rather than the GTV from the lung volume may reduce the apparent lung exposure, because normal lung within the PTV but outside the GTV will be excluded and may increase interinstitutional variations (because PTV margins may vary) [84]. Also, during treatment, there may be changes in GTV, with corresponding changes in normal tissue anatomy. Thus, plans defined on the basis of pre-RT imaging may not accurately reflect the degree of normal lung exposure.

The most widely used normal tissue complication probability model for RP is the Lyman–Kutcher–Burman (LKB) model, which has three parameters: a position parameter, TD_{50}; a steepness parameter, m; and the volume exponent, n [where n = 1, the model reverts to mean lung dose (MLD)] [85].

In a landmark publication, Graham et al. [86] described a close predictive value of V_{20} (volume of lung receiving 20 Gy) with the development of pneumonitis, the incidence raising substantially with V_{20} values of ≥40% lung volumes. Some data suggest that the lower lobes of the lung are more susceptible to pneumonitis compared with the upper lobes [87]. Bradley et al. [88], using RTOG 93-11 protocol data, designed a nomogram to predict RP and found the most predictive value was D15.

Kwa et al. [89] carried out a study of 540 patients from three centers with lung cancer, breast cancer, or lymphoma to evaluate factors involved in RP; the physical dose distribution was converted into a normalized TD (NTD) distribution using the linear quadratic model with an α/β ratio of 3.0 Gy (centers 1 and 5) or 2.5 Gy (center 3). The NTD is defined as the TD given in 2-Gy fractions, which is biologically equivalent to the actual treatment schedule. Lung volume was defined with CT, excluding GTV. The incidence at 4–16 Gy in the lung group of center 5 was 3% (2/62), and significantly ($p = 0.04$) lower than that of centers 2, 3, and 4, which was 13% (19/144). Barriger et al. [90], in a review of 167 patients with stage III non-small-cell lung cancer receiving concurrent cisplatinum and etoposide with modern 3D radiation techniques to 59.4 Gy, found that the only dosimetric parameter to correlate with grade ≥2 pneumonitis was MLD > 18 Gy.

In patients treated with accelerated partial breast irradiation, Recht et al. [91] reported four patients of 198 with pneumonitis when the volume of lung irradiated was larger. They recommended constraints for the ipsilateral lung of 3% for V_{20}, 10% for V_{10}, and 20% for V5.

As stated by Marks et al. [85], it is prudent to limit V_{20} to ≤30–35% and MLD to ≤20–23 Gy (with conventional fractionation) to reduce the risk of RP to ≤20% in definitively treated patients with non-small-cell lung cancer. After pneumonectomy or for mesothelioma, it is prudent to limit the V_5 to <60%, the V_{20} to <4–10%, and the MLD to <8 Gy [92]. Limiting the dose to the central airways to ≤80 Gy reduces the risk of bronchial stricture.

3.4.17 Esophagus

Esophagitis begins to develop with doses >20 Gy. DVH parameters describing cumulative dose >50 Gy have been identified as significantly correlated with acute esophagitis in several studies, although some have shown the strongest statistically significant correlations with esophagitis at lower doses (as low as V_{30}), perhaps owing to technique differences. In RTOG protocol 8311 with twice-daily hyperfractionated RT, grade 3 acute esophagitis was observed in 4–9% of patients receiving escalating doses from 60 Gy to 79.7 Gy. The risk of esophageal stricture ranges from 1% with 50 Gy to 5% with 60 Gy; the length of organ irradiated in some studies has been shown to correlate with incidence of late toxicity [94]. Maguire et al. noted that patients treated with ≥50 Gy to ≥32% volumes of the esophagus had increased late toxicity. Accelerated hyperfractionated irradiation (60 Gy in 30 fractions twice daily) increases the incidence of symptomatic acute esophagitis to 20–30% [95].

Concurrent chemotherapy increases both the severity and incidence of acute and late esophageal sequelae. Bradley et al. [96] reported that the area of the esophagus receiving doses >55 Gy, organ volume ≥60 Gy, and concurrent chemotherapy were associated with greater incidence of acute esophagitis. In RTOG protocol 0613, which delivers 60 Gy in three fractions with SBRT for stages I–II non-small-cell lung cancer to any point of the esophagus, is limited to a maximum of 27 Gy (9 Gy per fraction), representing 45% of the target dose.

3.4.18 Thoracic Cage

A very low incidence of rib necrosis/fracture is observed with ≤50 Gy, 3–5% with 60 Gy, and a greater incidence with higher doses. Large dose per fraction increases the risk; in a study by Overgaard [97] in patients irradiated postmastectomy to a dose of 51.3 Gy in 22 fractions (2.3 Gy/fraction) or 46.4 Gy in 12 fractions (3.9 Gy/fraction) twice weekly, the incidence of rib fracture was 6% vs 19%, respectively.

The increasing use of SBRT to treat early lung cancer has drawn attention to the importance of tolerance of the ribs to these radiation doses. Dunlap et al. [98] reported that the volume of chest wall (CW) (threshold 30 cm^3) receiving >30 Gy was predictive of severe pain and rib fracture. A 30% risk of developing severe CW toxicity correlated with a CW volume of 35 cm^3 receiving 30 Gy [99]. Andolino et al. [100], in a study of 347 patients with malignant lung and liver lesions, concluded that a dose of 50 Gy was the cutoff for maximum dose (D_{max}) to CW and rib above which there was a significant increase in the frequency of any grade of pain and fracture ($p = 0.03$ and $p = 0.025$, respectively). Mutter et al. [101], in a study of 126 patients treated with 40–60 Gy in three to five fractions noted that the volume of the CW receiving 30 Gy correlated with grade 2 chest pain was 70 cm^3.

3.4.19 Breast

A common sequela of breast irradiation is fibrosis, particularly in the boost volume. Mukesh et al. [102], reviewed the current literature and recorded dose–volume effect of breast irradiation. In the European Organisation for Research and Treatment of Cancer (EORTC) boost versus no boost trial, 5,318 patients with early breast cancer were randomized between a boost of 16 Gy with electrons or with iridium-192 implant (dose rate 0.5 Gy/h) versus no boost after whole-breast irradiation (WBI) (50 Gy). At 10 years, the incidence of moderate to severe breast fibrosis was 28.1% versus 13.2%, respectively ($p < 0.0001$) [103]. In that trial, patients with microscopically incomplete tumor excision were also randomized to either a boost of 10 Gy (126 patients) or 26 Gy (125 patients). The cumulative incidence of moderate/severe fibrosis for low-dose and high-dose boosts at 10 years was 24% and 54%, respectively. The boost volume for the group with complete excision around the tumor bed was a 1.5-cm margin compared with tumor bed plus 3-cm margin in the incomplete tumor excision group. This result demonstrated the impact of an increase in irradiated breast volume in doubling the risk of moderate/severe fibrosis for the same dose escalation of 16 Gy [104]. At the Royal Marsden Hospital and Gloucestershire Oncology Centre trial [105], 1,410 patients with early breast cancer were randomized into three WBI regimens: the control arm, 50 Gy in 25 fractions over 5 weeks; the two test arms, one with 39 Gy in 13 fractions over 5 weeks and the other with 42.9 Gy in 13 fractions over 5 weeks. The equivalent doses in 2-Gy fractions using an α/β ratio of 3.1 Gy for palpable breast induration were 46.7 Gy and 53.8 Gy for test arms 1 and 2, respectively. The risk of moderate to severe induration at 10 years for arms 1 and 2 was 27% and 51%, respectively.

Accelerated partial breast irradiation (APBI) has also been implicated in postradiation breast sequelae. The Christie group randomized 708 patients with breast cancer to APBI (40–42.5 Gy in eight fractions over 10 days with electrons) or WBI (40 Gy, 15 fractions, 21 days). Patients treated with APBI had significantly higher rates of marked breast fibrosis (14% vs. 5%) and telangiectasia (33% vs. 12%) when compared with WBI [106].

3.4.20 Stomach

Incidental gastric irradiation in 256 patients with testicular cancer treated with orthovoltage irradiation led to gastric ulceration in 6% of patients receiving 45–50 Gy, 10% with 50–60 Gy, and 38% with doses > 60 Gy. No injury was noted with <45 Gy. With megavoltage equipment, gastric ulceration sometimes associated with perforation has been observed in 10–15% of the patients [108]. Goldstein et al. [109] noted radiological abnormalities of the distal stomach in 8% (10/121) of women 1–25 months after 50 Gy to the para-aortic nodes for metastatic cervical cancer. The lesions were all ulcers in or near the pylorus; only two required surgical intervention. In addition, 1 of 52 men who received 40–50 Gy of para-aortic nodal RT for testicular tumors developed gastric outlet obstruction secondary to a pyloric ulcer 3 months later.

3.4.21 Small Intestine

Late radiation effects in the small bowel may be associated with fibrosis and intestinal obstruction, or perforation, which usually happens with doses >50Gy with conventional fractionation (3% incidence) The Uppsala University Rectal Cancer Study compared preoperative pelvic RT, 25.5 Gy delivered in five fractions, versus 60 Gy in 7–8 weeks of split-course postoperative RT, with a reduced field for the last 10 Gy. Some patients did not have RT. At a minimum follow-up of 5 years, a surgical or radiographic diagnosis of small-bowel obstruction was made in 5% of patients (14/255) after preoperative RT, 11% (14/127) after postoperative RT, and 6% (5/82) after surgery alone [110]. According to Kavanagh et al. [111], RT doses on the order of 45 Gy to the whole stomach are associated with late effects (primarily ulceration) in approximately 5–7% of patients. For SBRT, the volume of stomach receiving >22.5 Gy should be minimized and ideally constrained to <4% of the organ volume, or approximately 5 cc, with maximum point dose <30 Gy for three-fraction SBRT.

Concurrent chemotherapy adds to RT-induced acute small-bowel toxicity. In a gynecologic oncology group study, patients with cervical cancer who received 45 Gy pelvic RT alone experienced a 5% (9/186) rate of grades 3–4 gastrointestinal toxicity versus 14% (26/183) from RT plus weekly cisplatin (40 mg/m^2) [24].

3.4.22 Rectum

In irradiation of patients with gynecologic or prostate cancer, limited volumes of the rectum routinely receive doses ≥60 Gy. In a large retrospective study of cervical cancer performed at the Mallinckrodt Institute of Radiology, the incidence of severe rectal morbidity was 4% with combined external-beam radiation therapy (EBRT) and low-dose-rate (LDR) brachytherapy doses of ≤75 Gy versus 9% with higher doses [112]. In patients irradiated for localized prostate cancer, rectal morbidity (proctitis, ulceration) is lower when <25% of the rectal volume receives doses of ≤70 Gy [113]. In another study, rectal bleeding occurred in 1% of patients when V_{65} was <23% and increased to 10% with V_{65} of 28% [114]. In a study of 101 patients with prostate cancer treated with 3D-CRT or IMRT (mean dose 70–74 Gy) and evaluated with rectoscopy within 1 year after treatment, V_{60} and V_{70} were correlated with incidence of rectal telangiectasis and bleeding [115]. Valdagni et al. [116], in an analysis of 718 patients with prostate cancer treated with 70–80 Gy in 1.8-2 Gy fractions, used a nomogram to evaluate late effects. They found that the most significant predictor of grade 3 rectal bleeding was V_{75} Gy and for grades 2–3 fecal incontinence, it was V_{70} Gy. Patients who underwent abdominal surgery before pelvic RT had a lower value for grade 3 rectal bleeding (V_{70} Gy).

Androgen depravation therapy has been implicated in increased rectal radiation toxicity [117].

In 48 patients irradiated for localized prostate cancer (67.5–70 Gy in 2.25-2.5 Gy fractions, Smeenk et al. [118] delineated the external anal sphincter (EAS), the internal anal sphincter (IAS), the levator ani muscles (LAM), and the puborectalis muscle (PRM), calculated doses received by each, and correlated results with the incidence of fecal incontinence. Based on dose–effect curves, the authors recommend the following constraints for mean anorectal doses to decrease the risk of rectal urgency: 10 Gy to EAS, 30 Gy to IAS, 40 Gy to LAM, and 50 Gy to PRM.

Data on late rectal tolerance to hypofractionated stereotactic irradiation (36-40 cGy in five fractions) has not been reported in several sudies [119].

In an European randomized trial (FFCD 9203) patients were randomized to receive preoperative pelvis rt (45 Gy) with or without bolus 5FU/leucovorin. Increased grade 3+ toxicity was reported in the group treated with RT+5FU/leucovorin (14% vs 2%, respectively, P=0.00001) [107].

3.4.23 Liver

It has been observed that alterations in liver function tests are present when doses >18 Gy are delivered to the liver. Austin–Seymour et al. [120] suggested that hepatitis developed with doses >18 Gy to the entire liver and 30 Gy to one third; <30% of the liver tolerates doses up to 50 Gy. Kutcher and Burman [121], in a group of patients irradiated with 3D-CRT, noted that those who developed radiation hepatitis received a mean whole-liver dose of 37 Gy. With hypofractionated stereotactic RT to limited liver volumes, it has been reported that the incidence of hepatitis was 7.8% for a single fraction of 19.7 Gy and 6.6% for four fractions of 8.8 Gy [122].

3.4.24 Kidney

Tolerance of the kidneys to irradiation has been studied in children irradiated for Wilm's tumor and neuroblastoma and in adults treated for testicular tumors. It is widely held that radiation nephritis develops with doses >23Gy to two thirds or more of one kidney [123]. Cassady [124] pooled data on bilateral whole-kidney RT tolerance and confirmed a threshold dose for RT injury of 15 Gy, with a 5-year risk of 5% for whole-kidney dose of 18 Gy, and 50% risk for 28 Gy within 5 weeks of exposure. Some publications point out that the TD 5/5 for two thirds of the renal parenchyma is 30 Gy [125]. Cheng et al. [126] found a less steep dose response after total body irradiation (TBI) (median dose 12 Gy in six fractions twice daily). The dose associated with a 5% risk of kidney toxicity was 9.8 Gy. The addition of nephrotoxic drugs made the dose–response curve steeper.

3.4.25 Bladder

In patients irradiated for gynecological tumors and prostate or bladder cancer, the tolerance of the urinary bladder has been established: for partial organ at 80 Gy and for whole organ at 50 Gy. Marks et al. [127] estimated a clinical complication rate of 5–10% with 50 Gy given to the whole bladder in 2-Gy fractions. Similar toxicity has been observed with 60–65 Gy to partial bladder volumes. The RTOG 0415 [140] study of prostate cancer patients included a bladder dose–volume constraint of no more than 15% of the volume to receive >80 Gy, no more than 25% of the volume to receive >75 Gy, no more than 35% of the volume to receive >70 Gy, and no more than 50% of the volume to receive >65 Gy. Urethral tolerance has been estimated at 65–70 Gy with 2-Gy external irradiation. In 67 patients treated with hypofractionated stereotactic RT (36.25 Gy in five fractions), grades 2–3 bladder toxicity was 3–5% [119].

3.4.26 Sexual Function

In men, irradiation of the penile bulb to doses >52.5 Gy have been associated with a greater risk of erectile dysfunction [128, 129]. In contrast, van der Wielen et al. [130] did not find a correlation between dose to the crura or penile bulb and erectile dysfunction 2 years after irradiation for prostate cancer. The testicles are very sensitive to ionizing radiation. Oligospermia or aspermia has been observed with doses of ≥10 Gy, with recovery dependent upon TD and fractionation. Anserini et al. [131] reported azoospermia in 81/85 (85%) patients who received TB fractionated irradiation (and cyclophosphamide) to doses of 9.9 or 13.2 Gy.

The ovaries are also very sensitive to irradiation, the effects being closely related to patient age at the time of exposure (remaining oocytes). Wallace et al. [132], with a mathematical model, calculated that the effective radiation dose to induce immediate/permanent sterilization in women was 20.3 Gy at birth, 18.4 Gy at 10 years of age, 16.5 Gy at 20 years of age, and 14.3 Gy at 30 years of age. Doses to cause ovarian failure were calculated as 18.9 Gy, 16.9 Gy, 14.9 Gy, and 12 Gy, respectively, at the same ages. Sequelae of vaginal irradiation range from mucosal atrophy to fibrosis, necrosis, or fistulae. The upper vagina tolerates irradiation better than the lower vagina, with doses resulting in necrosis being >140 Gy versus >98 Gy, respectively [133, 134], with a combination of EBRT and HDR brachytherapy. Sorbe and Smeds [135] reported with LDR brachytherapy a higher incidence of vaginal shortening with increasing dose per fraction. The traditional LDR tolerance dose of 150 Gy was shown to yield nominal 11% and 4% grades 1, 2, and 3 sequelae, respectively. With high-dose-rate (HDR) brachytherapy in fractions >7 Gy, greater morbidity has been noted compared with lower dose per fraction [136]. Au and Grigsby [137] calculated that the projected HDR grade 3 tolerance varied from 25 Gy for one fraction to 57 Gy for six fractions in addition to 20-Gy EBRT for nominal 3–5% complication rates.

3.4.27 Femoral Head

Osteonecrosis is rarely seen with radiation doses <60 Gy. However, doses in the range of ≤50 Gy are recommended for the femoral head and neck [141]. A synopsis of dose constraints for various OAR published in the QUANTEC Report [2] is presented in Table 3.2.

Table 3.2 Dose/volume/outcome data for several organs following conventional fractionation (unless otherwise noted): QUANTEC Summary

Organ	Volume segment	Irradiation type (partial organ unless otherwise stated)	End point	Dose (Gy), or dose/ volume parameters	Rate (%)	Notes on dose/ volume parameters
Brain	Whole organ	3D-CRT	Symptomatic necrosis	D_{max} <60	<3	Data at 72 and 90 Gy, extrapolated from BED models
	Whole organ	3D-CRT	Symptomatic necrosis	D_{max} = 72	5	
	Whole organ	3D-CRT	Symptomatic necrosis	D_{max} = 90	10	
	Whole organ	SRS (single fraction)	Symptomatic necrosis	V_{12} <5–10 cc	<20	Rapid rise when V_{12} > 5–10 cc
Brain stem	Whole organ	Whole organ	Permanent cranial neuropathy or necrosis	D_{max} <54	<5	
	Whole organ	3D-CRT	Permanent cranial neuropathy or necrosis	D1–10 ≤59	<5	
	Whole organ	SRS (single fraction)	Permanent cranial neuropathy or necrosis	D_{max} <12.5	<5	For patients with acoustic tumors
Optic nerve/ chiasm	Whole organ	3D-CRT	Optic neuropathy	D_{max} <55	<3	Given the small size, 3D-CRT is often whole organ
	Whole organ	3D-CRT	Optic neuropathy	D_{max} 55–60	3–7	
	Whole organ	3D-CRT	Optic neuropathy	D_{max} >60	>7–20	
	Whole organ	SRS (single fraction)	Optic neuropathy	D_{max} <12	<10	
Spinal cord	Partial organ Thoracic	3D-CRT	Myelopathy	D_{max} = 50	00.02	Including full cord cross-section
	Partial organ	3D-CRT	Myelopathy	D_{max} = 60	6	
	Cervical					
	Partial organ	SRS (single fraction)	Myelopathy	D_{max} = 13	1	Partial cord cross-section irradiated
	Partial organ	SRS (hypofractionation)	Myelopathy	D_{max} = 20	1	3 fractions, partial cord cross-section irradiated
Cochlea (auditory)	Whole organ	3D-CRT	Sensory neural hearing loss	Mean dose ≤45	<30	Mean dose to cochlea, hearing at 4 kHz
	Whole organ	SRS (single fraction)	Sensory neural hearing loss	Prescription dose ≤14	<25	Serviceable hearing

Modified from Marks et al [2].

QUANTEC Quantitative Analysis of Normal Tissue Radiation Effects in the Clinic, *CRT* conformal radiotherapy, *SRS* stereotactic radiosurgery, *GTV* gross tumor volume, *RILD* radiation-induced liver disease, *RTOG* Radiation Therapy Oncology Group, *BED* biologically equivalent dose, *SBRT* stereotactic body RT, *FLT4* tyrosine protein kinase receptor FLT4.

(continued)

Table 3.2 Dose/volume/outcome data for several organs following conventional fractionation (unless otherwise noted): QUANTEC Summary

Organ	Volume segment	Irradiation type (partial organ unless otherwise stated)	End point	Dose (Gy), or dose/volume parameters	Rate (%)	Notes on dose/volume parameters
Parotid	Bilateral whole Parotid glands	3D-CRT	Long-term parotid salivary function reduced to <25% of pre-RT level	Mean dose <25	<20	For combined parotid glands
	Unilateral whole parotid gland	3D-CRT	Long-term parotid salivary function reduced to <25% of pre-RT level	Mean dose <25	<20	For single parotid gland. At least one parotid gland spared to <20 Gy
	Bilateral whole parotid glands	3D-CRT	Long-term parotid salivary function reduced to <25% of pre-RT level	Mean dose <39	<50	For combined parotid glands
Pharynx	Pharynx constrictor muscles	Whole organ	Symptomatic dysphagia and aspiration	Mean dose <50	<20	Based on Section B4 in paper
Larynx	Whole organ	3D-CRT	Vocal dysfunction	Dmax <66	<20	With chemotherapy, based on single study
	Whole organ	3D-CRT	Aspiration	Mean dose <50	<30	With chemotherapy, based on single study
	Whole organ	3D-CRT	Edema	Mean dose <44	<20	Without chemotherapy, based on single study in patients without larynx cancer
	Whole organ	3D-CRT	Edema	V_{50} <27%	<20	
Lung	Whole organ	3D-CRT	Symptomatic pneumonitis	$V_{20} \le 30\%$	<20	For combined lung. Gradual dose response
	Whole organ	3D-CRT	Symptomatic pneumonitis	Mean dose = 7	5	Excludes purposeful whole-lung irradiation
	Whole organ	3D-CRT	Symptomatic pneumonitis	Mean dose = 13	10	
	Whole organ	3D-CRT	Symptomatic pneumonitis	Mean dose = 20	20	
	Whole organ	3D-CRT	Symptomatic pneumonitis	Mean dose = 24	30	
	Whole organ	3D-CRT	Symptomatic pneumonitis	Mean dose = 27	40	

(continued)

Table 3.2 *(continued)*

Organ	Volume segment	Irradiation type (partial organ unless otherwise stated)	End point	Dose (Gy), or dose/volume parameters	Rate (%)	Notes on dose/volume parameters
Esophagus	Whole organ	3D-CRT	Grade ≥3 acute esophagitis	Mean dose <34	5–20	Based on RTOG and several studies
	Whole organ	3D-CRT	Grade ≥2 acute esophagitis	V_{35} <50%	<30	A variety of alternate threshold doses have been implicated. Appears to be a dose/volume response
	Whole organ	3D-CRT	Grade ≥2 acute esophagitis	V_{50} <40%	<30	
	Whole organ	3D-CRT	Grade ≥2 acute esophagitis	V_{70} <20%	<30	
Heart	Pericardium	3D-CRT	Pericarditis	Mean dose <26	<15	Based on single study
	Pericardium	3D-CRT	Pericarditis	V_{30} <46%	<15	
	Whole organ	3D-CRT	Long-term cardiac mortality	V_{25} <10%	<1	Overly safe risk estimate based on model predictions
Liver	Whole liver – GTV	3D-CRT or Whole organ	Classic RILD	Mean dose <30–32	<5	Excluding patients with pre-existing liver disease or hepatocellular carcinoma
	Whole liver – GTV	3D-CRT	Classic RILD	Mean dose <42	<50	
	Whole liver – GTV	3D-CRT or Whole organ	Classic RILD	Mean dose <28	<5	In patients with Child-Pugh A pre-existing liver disease or hepatocellular carcinoma, excluding hepatitis B reactivation as an endpoint
	Whole liver – GTV	3D-CRT	Classic RILD	Mean dose <36	<50	
	Whole liver –GTV	SBRT (hypofractionation)	Classic RILD	Mean dose <13–18	<5; <5	3 fractions, for primary liver cancer, 6 fractions, for primary liver cancer
	Whole liver – GTV	SBRT (hypofractionation)	Classic RILD	Mean dose <15–20	<5; <5	3 fractions, for liver metastases, 6 fractions, for liver metastases
	>700 cc of normal liver	SBRT (hypofractionation)	Classic RILD	D_{max} <15	<5	Based on critical volume, in 3–5 fractions

Table 3.2 Dose/volume/outcome data for several organs following conventional fractionation (unless otherwise noted): QUANTEC Summary

Organ	Volume segment	Irradiation type (partial organ unless otherwise stated)	Endpoint	Dose (Gy), or dose/ volume parameters	Rate (%)	Notes on dose/ volume parameters
Kidney	Bilateral whole kidney	Bilateral whole organ or 3D-CRT	Clinically relevant renal dysfunction	Mean dose <15–18	<5	
	Bilateral whole kidney	Bilateral whole organ	Clinically relevant renal dysfunction	Mean dose <28	<50	
	Bilateral whole kidney	3D-CRT	Clinically relevant renal dysfunction	V_{12} <55% V_{20} <32% V_{23} <30% V_{28} <20%	<5	For combined kidney
Stomach	Whole organ	Whole organ	Ulceration	D_{100} <45	<7	
Small bowel	Individual small-bowel loops	3D-CRT	Grade ≥ 3 acute toxicity§	V_{15} <120 cc	<10	Volume based on segmentation of the individual bowel loops, not the entire potential peritoneal space
	Entire potential space within peritoneal cavity	3D-CRT	Grade ≥ 3 acute toxicity§	V_{45} <195 cc	<10	Volume based on the entire potential space within the peritoneal cavity
Rectum	Whole organ	3D-CRT	Grade ≥ 2 late rectal toxicity, grade ≥ 3 late rectal toxicity	V_{50} <50%	<15; <10	Prostate cancer treatment
	Whole organ	3D-CRT	Grade ≥ 2 late rectal toxicity, grade ≥ 3 late rectal toxicity	V_{60} <35%	<15; <10	
	Whole organ	3D-CRT	Grade ≥ 2 late rectal toxicity, grade ≥ 3 late rectal toxicity	V_{65} <25%	<15; <10	
	Whole organ	3D-CRT	Grade ≥ 2 late rectal toxicity, grade ≥ 3 late rectal toxicity	V_{70} <20%	<15; <10	
	Whole organ	3D-CRT	Grade ≥ 2 late rectal toxicity, grade ≥ 3 late rectal toxicity	V_{75} <15%	<15; <10	

(continued)

Table 3.2 *(continued)*

Organ	Volume segment	Irradiation type (partial organ unless otherwise stated)	Endpoint	Dose (Gy), or dose/volume parameters	Rate (%)	Notes on dose/ volume parameters
Bladder	Whole organ	3D-CRT	Grade ≥ 3 late RTOG	D_{max} <65	<6	Bladder cancer treatment. Variations in bladder size/shape/location during RT hamper ability to generate accurate data
	Whole organ	3D-CRT	Grade ≥3 late RTOG	V_{65} ≤50% V_{70} ≤35% V_{75} ≤25% V_{80} ≤15%		Prostate cancer treatment. Based on RTOG 0415 recommendation
Penile bulb	Whole organ	3D-CRT	Severe erectile dysfunction	Mean dose to 95% of gland <50	<35	
	Whole organ	3D-CRT	Severe erectile dysfunction	D_{90} <50	<35	
	Whole organ	3D-CRT	Severe erectile dysfunction	D_{60-70} <70	<55	FLT4
Upper femora	Whole bone	Any	Fracture	45–50	5	

3.5 Stereotactic, Stereotactic Body, Hypofractionated Radiation Therapy Tolerance Dose Constraints

The unorthodox fractionation schedules mentioned previously are being used increasingly in clinical practice throughout the world. The major feature is the delivery of large radiation doses per fraction in a few fractions, resulting in a high BED. To minimize normal tissue toxicity, it is critical to accurately deliver the prescribed dose to the target, sparing sensitive adjacent structures. These techniques are used in the treatment of small primary tumors in the brain and lung; selected metastasis in the brain, lung, or liver; and in spinal or paraspinal sites. Hypofractionated schedules have been used in the irradiation of the whole breast or in accelerated partial breast irradiation and in the prostate.

Normal tissue tolerance doses with hypofractionation are quite different than with conventional irradiation, and available data are still immature. Particular attention should be paid to TD, fraction dose, time between fractions, and total treatment time. Various forms of image-guided and motion-management techniques should be carefully applied during treatment, for localization of the target and sensitive adjacent structures and strict quality assurance procedures must be followed to ensure safe delivery of the prescribed irradiation dose. Table 3.3 summarizes data on dose constraints published by various authors. Except in the setting of institutional review boards, phase I protocols and critical organ tolerance doses with hypofractionated schedules based on published data in peer-reviewed literature should be respected [138, 139].

Table 3.3 Summary of radiation threshold dose constraints for stereotactic and hypofractionated schedules published in the literature

Organ	Max critical volume	One fraction (Gy)	Three fractions (Gy)	Five fractions (Gy)	End point grade 3
Brain	100 %			20	Necrosis
Brain stem	<0.5 cc	10	18 (6 Gy/fx	23 (4.6 Gy/fx	Neuropathy
Spinal cord	< 1.2 cc	7	12.3(4.1 Gy/fx)	14.5 (2.9 Gy/fx	Myelopathy
Optic nerve	0.2 cc	08-ott	15	20	Neuropathy
Cochlea		10	17	23	Hearing loss
Larynx	4 cc	10		20	
Brachial plexus	3 cc	14	22.05	30	Neuropathy
Bronchus	< 4 cc	10	15 (5 Gy/fx)	16.5 (3.3 Gy)	
Lung	1,000 cc	07.04	10.5 (4 Gy/fx)	13.5 (2.7 Gy/fx)	Pneumonitis
Heart	< 15 cc	16	24 (8 Gy/fx)	32 (6 Gy/fx)	Pericarditis
Esophagus	< 5 cc	11.09	17	20	Stenosis
Rib	< 1 cc	22	28	35	Fracture
Stomach	< 10 cc	11	16.5 (5 Gy/fx)	18 (3.6 Gy/fx)	Ulceration
Duodenum	< 10 cc	9	11.04	12.05	Stenosis
Small bowel	< 5 cc	11.09	17.7 (5.9Gy/fx)	19.05	Stenosis
Colon/rectum	< 20 cc	14.03	16.8 (5.6 Gy/fx)	18.3 (3.6 Gy/fx)	Colitis Proctitis
Liver	< 700 cc	9	19 (6.4 Gy/fx)	21 (4.2 Gy/fx)	Liver function
Kidney	< 200 cc	08.04	16 (4 Gy/fx)	17.5 (3.5 Gy)	Renal function
Bladder	< 15 cc	11.04	16.8 (5.6 Gy)	18 (3.6 Gy/fx)	Cystitis
Penile bulb	< 3 cc	14	21.9 (7.3 Gy)	30 (6 Gy/fx)	Erectile dysfunction
Skin	< 10 cc	23	30 (10 Gy/fx)	36.5(7.3 Gy)	Ulceration
Femoral head	< 10 cc	14	21.9 (7.3 Gy)	30(6 Gy/fx)	Necrosis

Fx fraction.
Modified from Benedict et al. and Grimm et al. [138, 139].

Acknowledgement This chapter has been written with the contribution of Bahman Emami.

References

1. Jaffray DA, Lindsay PE, Brock KK et al (2010) Accurate accumulation of dose for improved understanding of radiation effects in normal tissue. Int J Radiat Oncol Bio Phys 76(Suppl 3):S135–S139
2. Marks LB, Ten Haken RK, Martel MK (2010) Guest editor's introduction to QUANTEC: A users guide. Int J Radiat Oncol Bio Phys 76(Suppl 3):S1–S2
3. Held KD, Willers H (2011) Molecular and cellular basis of radiation injury. in human radiation injury. In: Shrieve DC, Loeffler JS, eds. Wolters Kluwer Lippincott Williams & Wilkins pp 1–13
4. Coutard H (1932) Roentgen therapy of epitheliomas of the tonsillar region, hypopharynx and larynx from 1920 to 1926. AJR Am J Roengenol 28:313–331
5. Regaud C, Ferroux R (1927) Discordance des effects des Rayons X, d'une part dans la peau, l'autre part dans le testicle, par fractionnement de la dose: diminution de l'efficacie dans la peau, maintien de l'efficacite dans le testicule. C R Soc Biol 97:431–434
6. Paterson JR (1948) The treatment of malignant disease by radium x-rays, being a practice of radiotherapy. Edward Arnold, London
7. Strandqvist M (1944) Studien iiber die kumulative Wirkung der Rontgenstrahlen bei Fraktionierung. Erfahrungen aus demRadiumhemmet an 280 Haut und Lippenkarzinomen. Acta Radiologica Suppl (Stockh) 1944; 55:1–300
8. Ellis F (1969) Dose, time and fractionation: A clinical hypothesis. Clin Radiol 20:1–7
9. Ellis F (1971) Normal standard dose and the ret. Br J Radiol 44:101–108

10. Cohen L, Kerrick JE (1951) Estimation of biological dosage factors in clinical radiotherapy. Br. J. Cancer 5:180–194
11. Thames HD Jr, Withers HR, Peters LJ, Fletcher GH (1982) Changes in early and late radiation responses with altered dose fractionation: implications for dose-survival relationships. Int. J. Radial. Oncd. Biol. Phys 8:219–226
12. Dale RG (1985) The application of the linear-quadratic dose-effect equation to fractionated and protracted radiation. Brit. J. Radiol 58:515–528
13. Fowler JF, Tome WA, Fenwick JD et al (2004) A challenge to traditional radiation Oncology. Int J Radiat Oncol Bio Phys 60:1241–1256
14. Liu L, Bassano DA, Prasad SC et al (2003) The linear-quadratic model and fractionated stereotactic radiotherapy Int J Radiat Oncol Bio Phys 57:827–832
15. Niemierko A (1999) A generalized concept of equivalent uniform dose (EUD). Med Phys 26:1100
16. Emami B, Lyman J, Brown A et al (1991) Tolerance of normal tissue to therapeutic irradiation. Int J Radiat Oncol Biol Phys 21:109–122
17. Burman C, Kutcher GJ, Emami B et al (1991) Fitting of normal tissue tolerance data to an analytic function. *Int J Radiat Oncol Biol Phys* 21:123–135
18. Lyman JT (1985) Complication probability as assessed from dose-volume histograms. Radiat Res Suppl 8:S13–S19
19. Kutcher GJ, Burman C, Brewster L et al (1991) Histogram reduction method for calculating complication probabilities for three dimensional treatment planning evaluations. Int J Radiat Oncol Bio Phys 21:137–146
20. Trotti A, Colevas AD, Setser A et al (2003) CTCAE v3.0: Development of a comprehensive grading system for the adverse effects of cancer treatment. Sem Radiat Oncol 13:176–181
21. Bentzen SM, Constine LS, Deasy JO et al (2010) Quantitative Analyses of Normal Tissue Effects in the Clinic (QUANTEC): an introduction to the scientific issues. Int J Radiat Oncol Bio Phys 76(Suppl 3):S3–S9
22. Deasy JO, Bentzen SM, Jackson A et al (2010) Improving normal tissue probability models: The need to adopt a "data pooling" culture. Int J Radiat Oncol Bio Phys 76(Suppl 3):S151–154
23. Marks LB, Yorke ED, Jackson A et al. (2010) Use of Normal Complication Probability Models in the clinic. Int J Radiat Oncol Bio Phys 76(Suppl 3):S10–S19
24. Halperin EC, Perez CA, Brady LW (2008) The discipline of radiation oncology. In Halperin EC, Perez CA, Brady LW, eds. Principles and practice of radiation oncology, 5th edn. Wolters Kluwer Lippincott Williams & Wilkins, pp 26–28
25. McBride WH, Withers HR (2008) Biological basis of radiation therapy. In: Halperin EC, Perez CA, Brady LW, eds, Principles and practice of radiation oncology, 5th edn. Wolters Kluwer Lippicott Williams & Wilkins, p 10
26. Withers HR, Thames HD, Peters LJ (1983) A new isoeffect curve for change in dose per fraction. Radiother Oncol 1:187–191
27. Lee SP, Leu MY, Smathers JB et al (1995) Biologically effective dose distribution based on the linear quadratic model and its clinical relevance. Int J Radiat Oncol Bio Phys 33:372–389
28. Shaw E, Scott C, Souhami L, et al (2000) Single-dose radiosurgical treatment of recurrent previously irradiated primary brain tumors and brain metastases: Final report of RTOG protocol 90-05. *Int J Radiat Oncol Biol Phys* 47:291–298
29. Lawrence YR, Li XA, el Naqa I et al (2010) Radiation dose-volume effect in the brain. Int J Radiat Oncol Bio Phys 76(S3):S20–S27
30. De Angelis LM, Delattre JY, Posner JB (1989) Radiation-induced dementia in patients cured of brain metastasis. Neurology 39:789–796
31. Shih HA, Loeffler JS (2011) Hypothalamic–pituitary axis. In: Shrieve DC, Loeffler JS, eds, Human radiation injury. Wolters Kluwer Lippincott Williams & Wilkins, pp 180–189
32. Bhandare N, Kennedy L, Malyapa RS et al (2008) Hypopituitarism after radiotherapy for extracranial head and neck cancers. Head Neck 30: 1182–1192
33. Snyers A Janssens GORJ, Twickler MB (2009) Malignant tumors of the nasal cavity and paranasal sinuses: long-term outcome and morbidity with emphasis on hypothalamic-pituitary deficiency. Int J Radiat Oncol Bio Phys 73:1343–1351
34. Scicignano G, Losa M, del Vecchio A et al (2012) Dosimetric factors associated with pituitary function after gamma knife surgery (GKS) of pituitary adenomas. Radiother Oncol 104:119–124
35. Mayo C, Yorke E, Merchant TE (2010) Radiation associated brainstem injury. Int J Radiat Oncol Bio Phys 76(Suppl 3):S36–S41
36. Flickinger JC (2011) Cranial nerves. In: Shrieve DC, Loeffler JS, eds. Wolters Kluwer Lippincott Williams & Wilkins, p 210
37. Jackson A, Marks LB, Bentzen SM et al. (2010) The lessons of QUANTEC: Recommendations for reporting and gathering data on dose-volume dependencies of treatmen outcome. Int J Radiat Oncol Bio Phys 76(Suppl 3):S155–S160
38. Parsons JT, Fitzgerald C R, Hood C I et al (1983) The effects of irradiation on the eye and optic nerve. Int. J. Radiat. Oncol Biol Phys 1983; 9609–622
39. van den Bergh AC, Schoorl MA, Dullaart RP et al (2004) Lack of radiation optic neuropathy in 72 patients treated for pituitaryadenoma. J Neuroophthalmol 24:200–205
40. Hammer HM (1983) Optical chiasmal radionecrosis. Trans Ophthalmol Soc UK 103:208–211
41. Mayo C, Martel MK, Marks LB et al (2010) Radiation dose-volume effect of optic nerves and chiasm. Int J Radiat Oncol Bio Phys 76(Suppl 3):S28–S35
42. Tishler RB, Loeffler JS, Lunsford LD et al (1993) Tolerance of cranial nerves of the cavernous sinus to radiosurgery. Int J Radiat Oncol Bio Phys 27: 215–221
43. Bhandare N, Parsons JT, Bhatti MT, Mendenhall WM (2011) Optic nerve, eye and ocular adnexa. In:

Shrieve DC, Loeffler JS, eds. Wolters Kluwer Lippincott Williams & Wilkins, p 190
44. Marcus RB Jr, Million RR (1990) The incidence of myelitis after irradiation of the cervical spinal cord. Int J Radiat Oncol Bio Phys 19:3–8
45. Fowler JF, Bentzen SM, Bond SJ et al (2000) Clinical radiation doses for spinal cord: the 1998 international questionnaire. Radiother Oncol 55:295–300
46. Sahgal A, Wong CS, van der Kogel AJ (2011) Spinal cord in human radiation injury. In: Shrieve DC, Loeffler JS, eds. Wolters Kluwer Lippincott Williams & Wilkins, pp 190– 209
47. Kirpatrick JP, van der Kogel AJ, Schultheiss TE (2010) Radiation dose-volume effects in the spinal cord. Int J Radiat Oncol Bio Phys 76(Suppl 3):S42–S49
48. Daly ME, Luxton G, Choi CYH et al (2012) Normal tissue complication probability estimation by the Lyman-Kutcher-Burman method does not accurately predict spinal cord tolerance to stereotactic radiosurgery. Int J Radiat Oncol Bio Phys 82:2025–2032
49. Stoll BA, Andrews JT (1966) Radiation-induced peripheral neuropathy. Br Med J 1:834–837
50. Pierce SM, Recht A, Lingos T et al (1992) Long-term radiation complications following conservative surgery (CS) and radiation therapy (RT) in patients with early stage breast cancer. Int J Radiat Oncol Bio Phys 23:915–923
51. Guiou M, Hall WH, Jennelle R et al (2008) Prospective evaluation of dosimetric variables associated with brachial plexopathy after radiation therapy for head and neck cancer Int J Radiat Oncol Bio Phys 22(Suppl 1):S385
52. Fourquer JA, Fakiris AJ, Timmerman RD et al (2008) Brachial plexopathy (BP) from stereotactic body radiotherapy (SBRT) in early-stage NSCLC: dose limiting toxicity in apical tumor sites. Int J Radiat Oncol Bio Phys 72(Suppl 1):S36–S37
53. Henk JM, Whitelocke RA, Warrington AP et al (1993) Radiation dose to the lenses and cataract formation. Int J Radiat Oncol Bio Phys 25:815–820
54. Monroe AT, Bhandare N, Morris CG et al (2004) Preventing radiation retinopathy with hyperfractionation. Int J Radiat Oncol Bio Phys 60:S188
55. Bhandare N, Antonelli PJ, Morris CG et al (2007) Ototoxicity after radiotherapy for head and neck tumors. Int J Radiat Oncol Bio Phys 67:469–479
56. Chan SH, Ng WT, Kam KL et al (2009) Sensorineural hearing loss after treatment of nasopharyngeal carcinoma: a longitudinal analysis. Int J Rad iat Oncol Bio Phys 73:1335–1342
57. Bhandare N, Jackson A, Eisbruch A et al (2010) Radiation therapy and hearing loss. Int J Radiat Oncol Bio Phys 76(Suppl 3):S50–S57
58. Blanco AI, Chao KS, El Naqa I et al (2005) Dose-volume modeling of salivary function in patients with head and neck cancer receiving radiotherapy. Int J Radiat Oncol Bio Phys 62:1055–1069
59. Meirovitz A, Murdoch-Kinch CA, Schipper M et al (2006) Grading xerostomia by physicians or by patients after intensity modulated radiotherapy of head and neck cancer. Int J Radiat Oncol Bio Phys 66:445–453
60. Pow EH, Kwong DL, McMillan AS et al (2006) Xerostomia and quality of life after intensity modulated radiotherapy vs conventional radiotherapy for early-stage nasopharyngeal carcinoma: initial report on a randomized controlled clinical trial. Int J Radiat Oncol Bio Phys 66:981–991
61. Nutting CM, Morden JP, Harrington KJ et al (2011) Parotid-sparing intensity modulated versus conventional radiotherapy in head and neck cancer (PARSPORT): a phase 3 multicentre randomised controlled trial. Lancet Oncol 12:127–136
62. Munter MW, Hoffner S, Hof H et al (2007) Changes in salivary gland function after radiotherapy of head and neck tumors measured by quantitative pertechnetate scintigraphy: comparison of intensity modulated radiotherapy and conventional radiation therapy with and without Amifostine. Int J Radiat Oncol Bio Phys 67:651–659
63. Cannon DM, Lee NY (2008) Recurrence in region of spared parotid gland after definitive intensity-modulated radiation therapy for head and neck cancer. Int J Radiat Oncol Bio Phys 70:660–665
64. Wang ZH, Chao Y, Zhang ZY et al (2011) Impact of salivary gland dosimetry on post-IMRT recovery of saliva output and xerostomia grade for head and neck cancer patients treated with or without contralateral submaxillary gland sparing: a longitudinal study. Int J Radiat Oncol Bio Phys 81:1479–1487
65. Cooper JS, Fu K, Marks J (1995) Late effects of radiation therapy in the head and neck region. Int J Radiat Oncol Bio Phys 31:1141–1164
66. Dijkstra PU, Huisman PM, Roodenburg JL (2006) Criteria for trismus in head and neck oncology. Int J Oral Maxillofac Surg 35:337–342
67. Wang CJ, Huang EY, Hsu HC et al (2005) The degree and time-course assessment of radiation-induced trismus after radiotherapy for nasopharyngeal carcinoma. Laryngoscope 115:1458–1460
68. Cox JD, Pajak TF, Marcial VA et al (1991) ASTRO PLENARY: Interfraction interval is a major determinant of late effects, with hyperfractionated Radiation Therapy in carcinomas of the upper respiratory and digestive tracts: Results from RTOG protocol 8313. Int J Radiat Oncol Bio Phys 1:1191-1195
69. Sanguineti G, Adapala P, Endres EJ et al (2007) Dosimetric predictors of laryngeal edema. *Int J Radiat Oncol Biol Phys* 68:741–749
70. Bae JS, Roh J-L, Lee, S-W et al (2012) Laryngeal edema after radiotherapy in patients with squamous cell carcinoma of the larynx and pharynx. Oral Oncol 10:1016–1023
71. Rancati T, Schwarz M, Allen AM et al (2010) Radiation dose-volume effect in the larynx and pharynx. Int J Radiat Oncol Bio Phys 76(Suppl 3):S64–S69
72. Eisbruch A, Lyden T, Bradford CR et al (2002) Objective assessment of swallowing dysfunction and aspiration after radiation concurrent with chemotherapy

for head and neck cancer. Int J Radiat Oncol Bio Phys 53:23–28
73. Feng FY, Kim HM, Lynden TH et al (2007) Intensity modulated radiotherapy of head and neck cancer aiming to reduce dysphagia: early dose-effect relationships for the swallowing structures. Int J Radiat Oncol Bio Phys 68:1289–1298
74. Teguh DN, Levendag PC, Noever I et al (2008) Treatment techniques and site considerations regarding dysphagia-related quality of life in cancer of the oropharynx and nasopharynx. Int J Radiat Oncol Bio Phys 72:1119–11127
75. Caglar HB, Tishler, RB, Othus M et al (2008) Dose to larynx predicts for swallowing complications after intensity modulated radiotherapy. Int J Radiat Oncol Bio Phys 72:1110–1118
76. Caudell JJ, Schaner PE, Desmond RA et al (2010) Dosimetric factors associated with long-term dysphagia after definitive radiation therapy for squamous cell carcinoma of the head and neck. Int J Radiat Oncol Bio Phys 76:403–409
77. Alkan S, Baylancicek S, Ciftcic M et al (2008) Thyroid dysfunction after combined therapy for laryngeal cancer: a prospective study. Otolaryngol Head Neck Surg 139:787–791
78. Diaz R, Jaboin JJ, Morales-Paliza M et al (2010) Hypothyroidism as a consequence of intensity-modulated radiotherapy with concurrent taxane-based chemotherapy for locally advanced head and neck cancer. Int J Radiat Oncol Bio Phys 77:468–476
79. Bhandare N, Kennedy L, Malyapa RS et al (2007) Primary and central hypothyroidism after radiotherapy for head and neck tumors. Int J Radiat Oncolo Bio Phys 68:1131–1139
80. Swerdlow AJ, Higgins CD, Smith P et al (2007) Myocardial infarction mortality risk after treatment for Hodgkin disease: A collaborative British cohort study. J Natl Cancer Inst 99:206–214
81. Wei X, Liu HH, Tucker SL et al (2008) Risk factors for pericardial effusion in inoperable esophageal cancer patients treated with definitive chemoradiation therapy. Int J Radiat Oncol Biol Phys 70:707–714
82. Martel MK, Sahijdak WM, Ten Haken RK et al (1998) Fraction size and dose parameters related to the incidence of pericardial effusions. Int J Radiat Oncol Biol Phys. 40:155–161
83. Borger JH, Hooning MJ, Boersma LJ et al (2007) Cardiotoxic effects of tangential breast irradiation in early breast cancer patients: the role of irradiated heart volume. Int J Radiat Oncol Bio Phys 69:1131–1138
84. Feng M, Moran JM, Koelling T et al (2011) Development and validation of a heart atlas to study cardiac exposure to radiation following treatment for breast cancer. Int J Radiat Oncol Bio Phys 79:10–18
85. Marks LB, Bentzen SM, Deasy JO et al (2010) Radiation dose-volume effect in the lung. Int J Radiat Oncol Bio Phys 76(Suppl 3):S70–S76
86. Graham MV, Purdy JA, Emami B et al (1999) Clinical dose-volume histogram analysis for pneumonitis after 3D treatment for non-small cell lung cancer (NSCLC). Int J Radiat Oncol Biol Phys 45:323–329
87. Gomez DR, MD, Tucker SL, Martel MK et al (2012) Predictors of High-grade Esophagitis After Definitive Three-dimensional Conformal Therapy, Intensity-modulated Radiation Therapy, or Proton Beam Therapy for Non-small cell Lung Cancer. Int J Radiat Ocol Bio Phys 84:1010-1016
88. Bradley JD, Hope A, El Naqa I et al (2007) A nomogram to predict radiation pneumonitis derived from a combined analysis of RTOG 93-11 and institutional data. Int J Radiat Oncol Bio Phys 69:985–992
89. Kwa SLS, Lebesque JV, Theuws JCM et al (1998) Radiation pneumonitis as a function of mean lung dose: an analysis of pooled data of 540 patients. Int J Radiat Oncol Bio Phys 42:1-9
90. Barriger RB, Fakiris AJ, Hanna N et al (2010) Dose–volume analysis of radiation pneumonitis in non-small-cell lung cancer patients treated with concurrent cisplatinum and etoposide with or without consolidation docetaxel. Int J Radiat Oncol Biol Phys 78:1381–1386
91. Recht A, Ancukiewicz M, Liu X et al (2008) Lung Dose-Volume Parameters and the Risk of Pneumonitis for Patients Treated with Accelerated Partial-Breast Irradiation (APBI) using 3D Conformal Radiotherapy (3D-CR). Int J Radiat Oncol Bio Phys 72:S4-S5
92. Miles EF, Larrier NA, Kelsey CR et al (2008) Intensity-modulated radiotherapy for resected mesothelioma: The Duke experience. Int J Radiat Oncol Biol Phys 71:1143–1150
93. Marks, JE, Baglan RJ, Prasad, SC et al. (1981) Effects of radiation on parotid salivary function. Int. J.Radiat. Oncol. Biol. Phys 7:1013–1019
94. Maguire PD, Sibley GS, Zhou S-M et al (1999) Clinical predictors of radiation-induced esophageal toxicity. Int J Radiat Oncol Bio Phys 45:97-103
95. Saunders M, Dische S, Barrett A et al (1997) Continuous hyperfractionated accelerated radiotherapy (CHART) versus conventional radiotherapy in non-small-cell lung cancer: A randomized multicentre trial. Lancet 350:161–165
96. Bradley JD, Deasy JO, Bentzen S et al (2004) Dosimetric correlates for acute esophagitis in patients treated with radiotherapy for lung cancer. Int J Radiat Oncol Bio Phys 58:1106–1113
97. Overgaard M (1988) Spontaneous radiation-induced rib fractures in breast cancer patients treated with postmastectomy irradiation. A clinical radiobiological analysis of the influence of fraction size and dose-response relationship in late bone damage. Acta Oncol 27;117–122
98. Dunlap NE, Biederman GB, Yang W et al (2008) Chest wall volume receiving more than 30 Gy predicts risk of severe pain and/or rib fracture following lung SBRT. Int J Radiat Oncol Bio Phys 72(Suppl 1):S36
99. Dunlap NE, Cai J, Biederman GB et al (2010) Chest wall volume receiving >30 Gy predicts risk of severe pain and/or rib fracture after lung stereotactic

body radiotherapy. Int J Radiat Oncol Bio Phys 76:796–801

100. Andolino DL, Forquer JA, Henderson MA et al (2011) Chest wall toxicity after stereotactic body radiotherapy for malignant lesions of the lung and liver. Int J Radiat Oncol Bio Phys 80:692–697
101. Mutter RW, Liu F, Abreu A et al (2012) Dose-volume parameters predict for the development of chest wall pain after stereotactic body radiation for lung cancer. Int J Radiat Oncol Bio Phys 82:1783–1790
102. Mukesh M, Harris E, Jena R et al (2012) Relationship between irradiated breast volume and late normal tissue complications: A systematic review. Radiother Oncol 104:1–110
103. Bartelink H, Horiot JC, Poortmans PM et al (2007) Impact of a higher radiation dose on local control and survival in breast-conserving therapy of early breast cancer: 10-year results of the randomized boost versus no boost EORTC 22881-10882 trial. J Clin Oncol 25:3259–3265.
104. Poortmans PM, Collette L, Horiot JC et al (2009) Impact of the boost dose of 10 Gy versus 26 Gy in patients with early stage breast cancer after a microscopically incomplete lumpectomy: 10-year results of the randomised EORTC boost trial. Radiother Oncol 90:80–85
105. Yarnold J, Ashton A, Bliss J et al (2005) Fractionation sensitivity and dose response of late adverse effects in the breast after radiotherapy for early breast cancer: long-term results of a randomised trial. Radiother Oncol 75:9–17
106. Ribeiro GG, Magee B, Swindell R Et al (1993) The Christie Hospital Breast Conservation Trial: an update at 8 years from inception. Clin Oncol 5:278–283.
107. Conroy T, Bonnetain F, Chapet O et al (2004) Preoperative radiotherapy (RT) + 5FU/folinic acid (FA) in T3,4 rectal cancers: preliminary results of the FFCD 9203 randomized trial. Proc ASCO 22:247
108. Czito BG, Willet CG (2011) Stomach in human radiation injury. In: Shrieve DC, Loeffler JS, eds. Wolters Kluwer Lippincott Williams & Wilkins, pp 444–452
109. Goldstein HM, Rogers LF, Fletcher GH, Dodd GD (1975) Radiological manifestations of radiation-induced injury to the normal upper gastrointestinal tract. Radiology 117:135–140
110. Frykholm GJ, Glimelius B, Pahlman (1993). Preoperative or postoperative irradiation in adenocarcinoma of the rectum: Final treatment results of a randomized trial and an evaluation of late secondary effects. Dis Colon Rectum. 36:564–572
111. Kavanagh BD, Pan CC, Dawson LA et al (2010) Radiation dose-volume effect in the stomach and small bowel. Int J Radiat Oncol Bio Phys 76(Suppl 3):S101–S107
112. Perez CA, Grigsby PW, Lockett MA et al (1999) Radiation therapy morbidity in carcinoma of the uterine cervix : dosimetric and clinical correlation. Int J Radiat Oncol Bio Phys 44:855–866
113. Pollack A, Zagars GK, Smith LG et a (2000). Preliminary results of a randomized radiotherapy dose-escalation study comparing 70 Gy with 78 Gy for prostate cancer. J Clin Oncol 18:3904–3911
114. Peeters ST, Lebesque JV, Heemsberg WD et al (2006) Localized volume effects for late rectal and anal toxicity after radiotherapy for prostate cancer. Int J Radiat Oncol Bio Phys 64:1151–1161
115. Ippolito E, Deodato F, Macchia G et al (2012) Early radiation-induced mucosal changes evaluated by proctoscopy: Predictive role of dosimetric parameters. Radiother Oncol 104:103–108
116. Valdagni R, Kattan MW, Rancati T et al (2012) Is it time to tailor the prediction of radio-induced toxicity in prostate cancer patients? Building the first nomograms for late rectal syndrome. Int J Radiat Oncol Bio Phys 82:1957–1966
117. Liu M, Pickles T, Agranovich A et al (2004) Impact of neoadjuvant androgen ablation and other factors on late toxicity after external beam prostate radiotherapy. Int J Radiat Oncol Biol Phys;58:59–67
118. Smeenk RJ, Hoffman AL, Hopman W PM et al (2012) Dose-effect relationship for individual pelvic floor muscles and anorectal complaints after prostate radiotherapy. Int J Radiat Oncol Bio Phys 83:636–644
119. King CR, Brooks JD, Gill H et al (2012) Long-term outcomes from a prospective trial of stereotactic body radiotherapy for low-risk prostate cancer. Int J Radiat Oncol Biol Phys 82(2):877–82.
120. Austin-Seymour MM, Chen GTY, Castro J et al (1986) Dose volume histogram analysis of liver radiation tolerance. Int. J. Radiat. Oncol. Biol. Phys 12:31–35
121. Kutcher GJ, Burman C (1989) Calculation of complication probability factors for non-uniform normal tissue irradiation: the effective volume method. Int J Radiat Oncol Bio Phys 16: 1623–1630
122. Tai A, Erickson B, Li XA (2009) Extrapolation of normal tissue complication probability for different fractionations in liver irradiation. Int J Radiat Oncol Bio Phys 74:283–289
123. Flentje M, Hensley F, Gademann G et al (1993) Renal tolerance to nonhomogeneous irradiation : comparison of obderved effects to predictions of normal tissue complication probability from different biophysical models. Int J Radiat Oncol Bio Phys 27:25–30
124. Cassady JR (1995) Clinical radiation nephropathy. Int J Radiat Oncol Biol Phys. 31:1249–1256
125. Willet CG, Tepper JE, Orlow EL et al (1986) Renal complications secondary to radiation treatment of upper abdominal malignancies. Int J Radiat Oncol Bio Phys 12:1601–1604
126. Cheng J, Schultheiss T, Wong J (2008) Impact of drug therapy, radiation dose and dose rate on renal toxicity following bone marrow transplantation. Int

J Radiat Oncol Biol Phys. 71:436–443

127. Marks LB, Carroll PR, Dugan TC et al (1995) The response of the urinary bladder, urethra and ureter to radiation and chemotherapy. Int J Radiat Oncol Bio Phys 31:1257–1280
128. Fisch BM, Pickett B, Weinberg V et al (2001) Dose of radiation received by the bulb of the penis correlates with risk of impotence after three-dimensional radiotherapy for prostate cancer. Urology 57:955–959
129. Roach M, Winter K, Michalski JM et al (2004) Penile bulb dose and impotence after three-dimensional conformal radiotherapy for prostate cancer on RTOG 9406: findings from a prospective multi-institutional phase I/II dose escalation study. Int J Radiat Oncol Bio Phys 60:1351–1356
130. Van der Wielen GJ, Hoogeman MS, Dohle GR et al (2008) Dose-volume parameters of the corpora cavernosa do not correlate with erectile dysfunction after external beam radiotherapy for prostate cancer: results from dose-escalation trial. Int J Radiat Oncol Bio Phys 71:795–800
131. Anserini P, Chiodi S, Spinelli S et al (2002) Semen analysis following allogeneic bone marrow transplantation. Additional data for evidence based counseling. Bone Marrow Transplant 30:447–451
132. Wallace WH, Thomson AB, Saran F et al (2004) Predicting age of ovarian failure after radiation to a field that includes the ovaries. Int J Radiat Oncol Bio Phys 33:637–659
133 Hintz BL, Kagan AR, Chan P et al (1980) Radiation tolerance of the vaginal mucosa. Int J Radiat Oncol Bio Phys 6:711-716
134 Eifel PJ, Levenback C, Wharton JT et al (1995) Time course and incidence of late conplications in patients treated with radiation therapy for FIGO stage IB carcinoma of the uterine cervix. It J Radiat Oncol Bio Phys 195:1289-1300
135 Sorbe BG and Smeds AC (1990) Postoperative vaginal irradiation with high dose rate afterloading technique in endometrial carcinoma stage I. Int J Radiat Oncol Bio Phys 18:305-314
136. Orton CG, Seyedsadr M, Somnay A (1991) Comparison of high and low dose rate remote afterloading for cervix cancer and the importance of fractionation. Int J Radiat Oncol Bio Phys 21:1425–1434
137. Au SP, Grigsby PW (2003) The irradiation tolerance dose of the proximal vagina. Radiother Oncol 67:77–85
138. Benedict SH, Yenice KM, Followill D et al (2010) Stereotactic body radiation therapy: The report of AAPM Task Group 101, Med Phys 37: 4078–4101
139. Grimm J, LaCouture T, Croce R et al (2011) Dose tolerance limits and dose volume histogram evaluation for stereotactic body radiotherapy. J Applied Clin Med Phys 12:267–292
140. RTOG Protocol 0415 (2009) A Phase III Randomized Study of Hypofractionated 3DCRT/IMRT versus Conventionally Fractionated 3DCRT/IMRT in Patients Treated for Favorable-Risk Prostate Cancer. Lee WR, Chairman, closed to accrual Dec 11, 2009
141. Grigsby PW, Roberts HL, Perez CA (1995) Femoral neck fracture following groin irradiation. Int J Radiat Oncol Bio Phys 32:63-67

Chapter 4
Volumetric Acquisition: Technical Notes

Identifying clinical volumes in radiotherapy is based on planning computed tomography (CT) [1]. Morphological imaging can be associated with functional imaging providing more precise details in defining volumes (Fig. 4.1 a–c).

Integrated imaging must have the following characteristics:

- provide an exact representation of normal anatomy;
- provide excellent contrast resolution between the tumor and normal tissues at risk;
- assess tumor heterogeneity in terms of metabolism, hypoxia, and cell proliferation;
- allow method reproducibility with small inter- and intraobserver variability;
- provide balanced cost–benefit ratio.

The two most commonly used imaging methods aimed at integrating volume-planning radiotherapy are CT and MRI [2, 3]. This combination is particularly useful because MRI improves delineation of some critical organs, such as structures of the central and peripheral nervous system (Fig. 4.2), anatomical-district otorhinolaryngology (Fig. 4.3), heart (Fig. 4.4), and urogenital diaphragm (Fig. 4.5).

Multiplanar acquisitions of both methods represent a further valuable support for accuracy in the clinical delineation of volumes of interest for higher detailed information in the 3D morphological evaluation [4] (Fig. 4.6).

Employing integrated imaging, particular attention should be paid to distortion of images and contours – typical of MRI – as a potential source of error for the radiation oncologist, especially if coregistration with CT planning is performed. It is therefore important to apply a rigorous methodology using integrated imaging in radiotherapy planning. The fusion of images includes two distinct and successive phases: (1) coregistration, and (2) fusion. In the coregistration phase, two images are matched in the same coordinate system. The information of mutual correspondence between one or more normal or pathological anatomical structures already provide an increase in information content compared with the same images considered separately and can optimize volume definition. The next step is fusion, which creates a new image containing information and aspects of both original images. Image fusion must be obtained by using several algorithms [5]:

- Landmark based, in which fusion is performed using at least three points identified on the reference image; these points are matched with three similar points of the complementary image;
- Interactive, in which fusion is based on the operator's ability to empirically recognize anatomical structures and establish a match between the main and complementary images;
- Frame based, in which correspondence is realized by acquiring both main and complementary images using a stereotactic device; therefore, the images have common external common landmarks;
- Contour base, which consists of image contouring and then on their fusion;

G. Ausili Cefaro, D. Genovesi, C.A. Perez, *Delineating Organs at Risk in Radiation Therapy*,
DOI: 10.1007/978-88-470-5257-4_4, © Springer-Verlag Italia 2013

Fig. 4.1a,b Brain integrated imaging

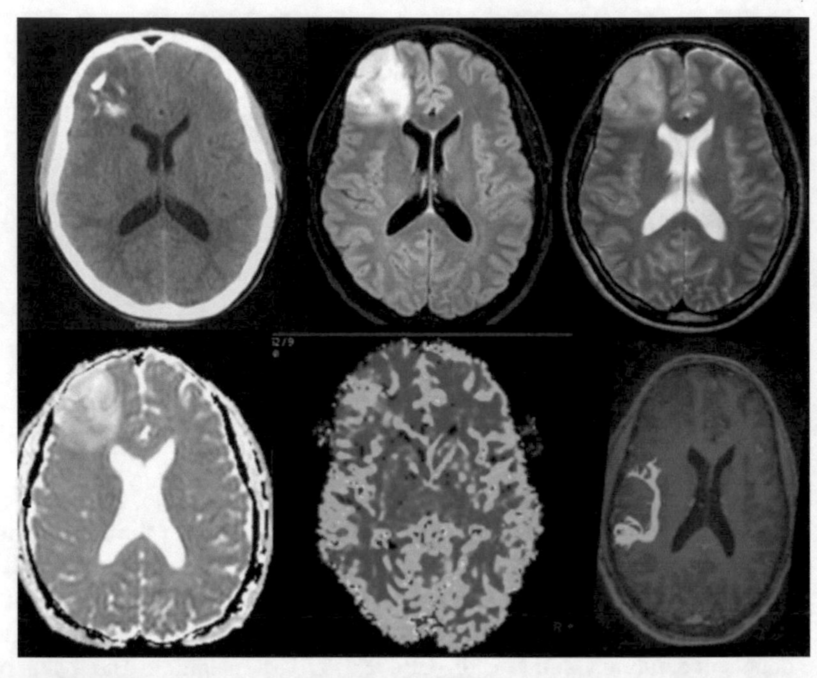

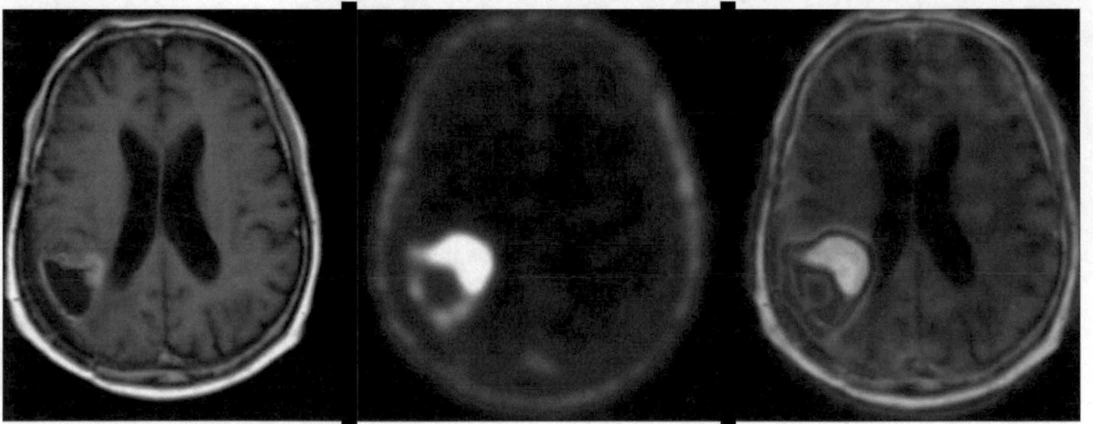

- Chamfer matching, which is a complex algorithm to determine structure based on both different threshold signal and automatic contouring; this method is useful for very different images, such as those provided by CT and PET;
- Volume matching, which is used for image fusion between CT and MRI of the central nervous system;
- Warping, another complex and impact algorithm; it permits obtaining fusion images, selectively deforming the images and thus increasing correspondence [5].

It is highly recommended, however, that regardless of the type of imaging employed for contouring (CT, the gold standard of radiotherapy planning; MRI; PET), a multidisciplinary team consisting of the radiotherapist, the radiologist, and a nuclear physician should be established. This multidisciplinary team is crucial in order to acquire an adequate level of expertise in the correct methodology of volumetric acquisitions rather than interpretation of individual imaging modalities alone.

Patient positioning and immobilization at the time of volume acquisition are other critical as-

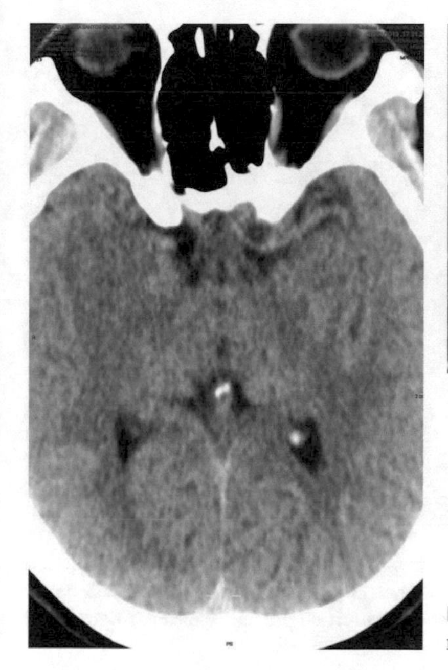

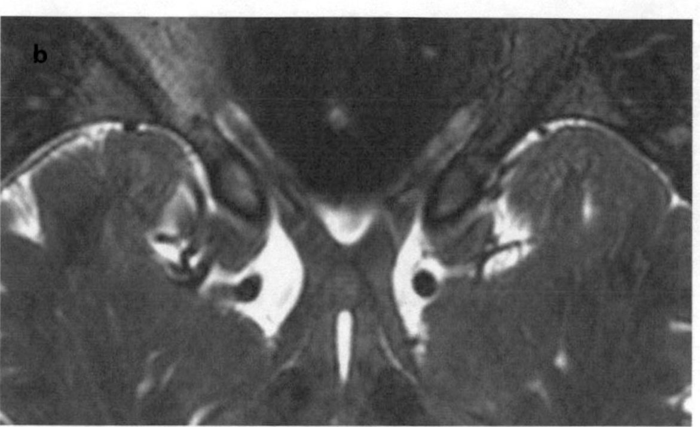

Fig. 4.2a,b Integrated imaging computed tomography–magnetic resonance imaging (CT-MRI) for optic chiasm delineation

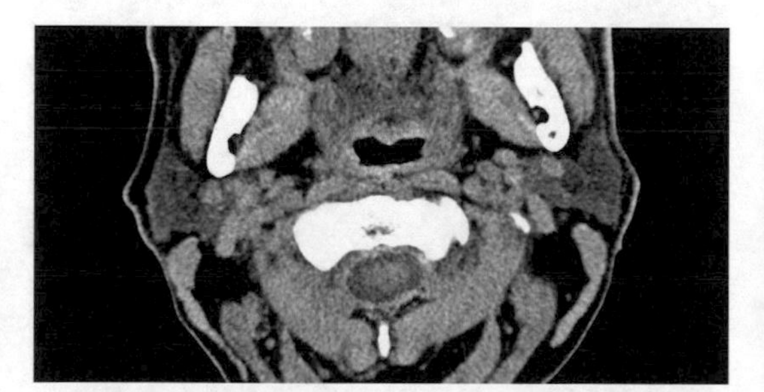

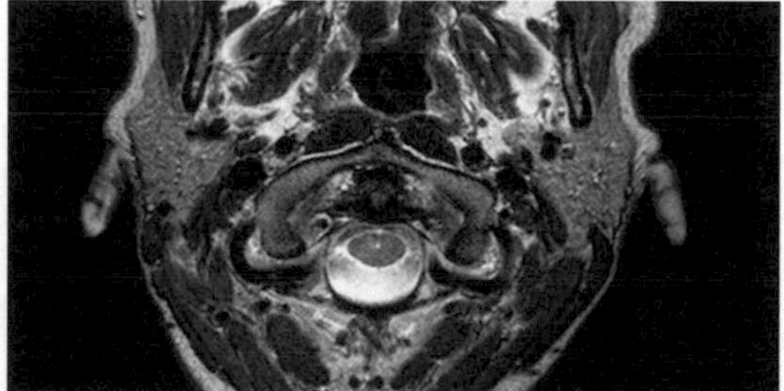

Fig. 4.3a,b Integrated imaging computed tomography–magnetic resonance imaging (CT-MRI) for parotid delineation

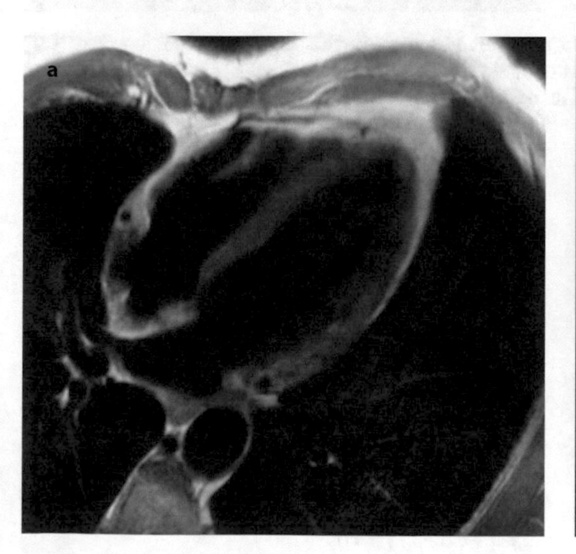

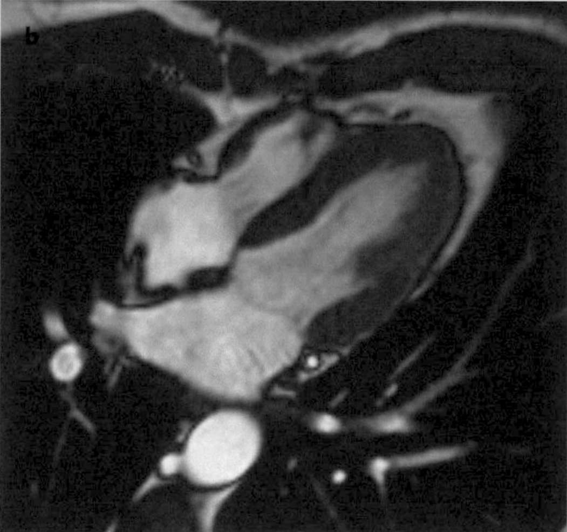

Fig. 4.4a,b Visualization of the heart using magnetic resonance imaging (MRI)

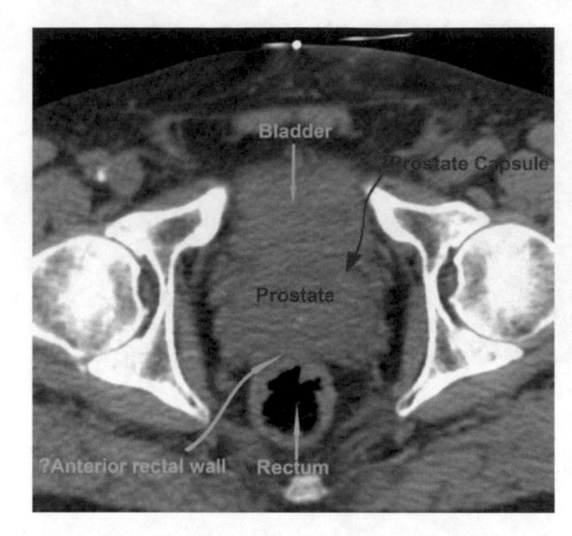

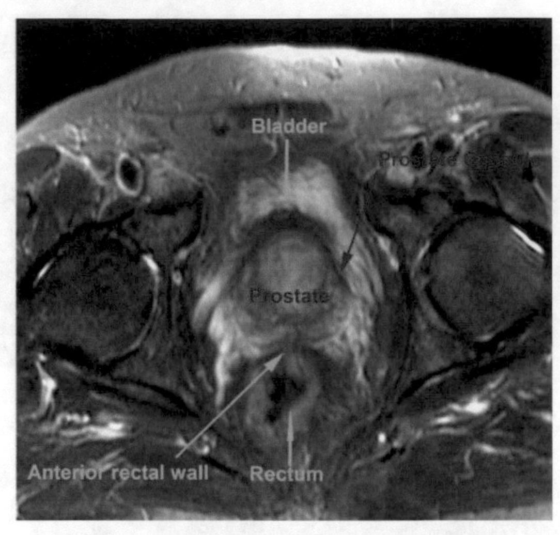

Fig. 4.5a,b Comparison of transaxial coregistered views of the pelvis for prostate radiotherapy with: **a** computed tomography (CT) and **b** magnetic resonance imaging (MRI) in the same patient. Boundaries of the prostate are better visualized using MRI than CT, particularly the anterior rectal wall/rectovesical fascia and prostate capsule. Reproduced with permission from [4]

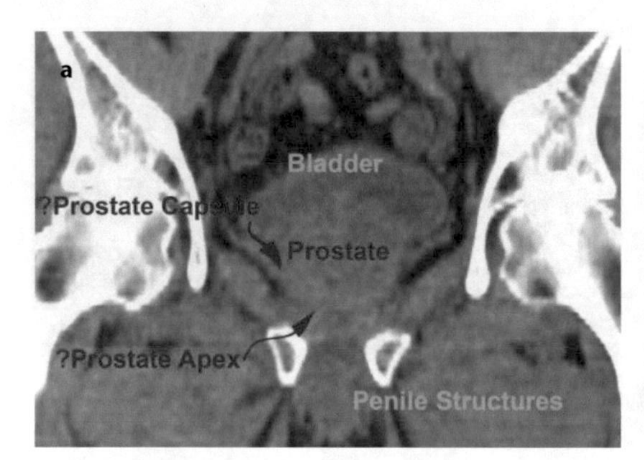

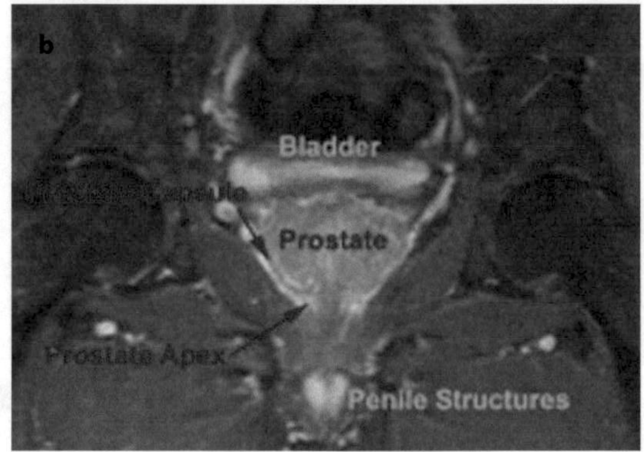

Fig. 4.6a,b Comparison of coronal views of the pelvis for prostate radiotherapy with: **a** computed tomography (CT) and **b** magnetic resonance imaging (MRI) in the same patient. Definition of prostate-gland boundaries and adjacent organs at risk is confirmed to be better visualized on MRI than CT. Reproduced with permission from [4]

pects in volume-planning radiotherapy, because they can ensure geometric reproducibility.

In radiation planning on the brain and head and neck districts, the patient assumes a supine position and, as a rule, head support must be used to obtain neutral or, if necessary, hyperextension of the neck. The head, neck, and shoulders of the patient are usually immobilized in this position using a thermoplastic mask locked to a base of a support on the treatment couch. Furthermore, depending on the type of treatment to be performed, other systems of immobilization may also be used (i.e., intraoral stent). The patient's arms are arranged alongside the body but crossing the hands on the body of the sternum may be considered to lower the shoulders and reduce the presence of artifacts on CT images caused by beam hardening [1, 6].

When irradiating the mediastinal region, the patient is placed supine with both arms above the head. Customized positioning systems, immobilization, and procedures of controlled breathing are desirable to reduce the uncertainty of geometric treatment (i.e., vacuum system, T-bar device; Perspex cast) [1, 6].

For treatments extended to the upper abdomen, patient position is usually supine, and it is advantageous to use immobilization systems (i.e., cast or vacuum system). Furthermore, to allow better positioning of fields treatment, the arms should be elevated above the head. Even in the abdominal area, the use of procedures of controlled breathing is recommended [1, 6].

In some treatments of the pelvic region, in addition to placing the patient in the supine position (cancer of the anus and gynecological), the prone position may be recommended (rectal cancer).

Irradiation of prostate tumors is carried out with the patient in the supine position in most cases; nevertheless, the prone position is described in some conditions.

Furthermore, it is possible to use a support under the patient's knees to allow good relaxation of the back, and a specific media device can also be used to reduce changes in feet position and improve accuracy of the setup [1, 6].

In the usual approach to stereotactic radiosurgery, a mechanical device, called the stereotactic frame, is attached to the patient's body. When fixed to the skull, the frame can also immobilize the patient. Less-invasive mechanical, optical, and radiographic methods can be reliably reapplied for fractionated treatment [7]. Such optical and radiographic methods are designated as frameless stereotactic techniques. The stereotactic devices in these cases may include a bite block that can be located in place with a stereo-optical camera [8, 9], a stereopair of kilovoltage fluoroscopy units [10], or a combination of various imaging and tracking methods [11]. In each case, the basic purpose of the stereotactic device is to provide an accurate coordinate system within which to direct a radiation beam to the target.

Radiation oncologists and Radiologists at the University of Chieti developed an interdisciplinary methodology that led to the formulation of practical advice in planning CT. It is recommended that planning CT is performed with spiral acquisition in a single layer, as it can provide a higher resolution and an information volume higher than the acquisition sequential [12–14]. It is usually performed with the patient breathing freely and almost always without intravenous contrast administration. In the presence of peripheral lung cancers, it is possible to consider optimizing the above procedure by acquiring scans with slow-mode CT (4 s per scan), obtained with the patient breathing quietly, or by using the 4D-CT technique [15].

CT provides registration of two topograms: anteroposterior and laterolateral.

For the four main anatomical regions considered in our experience, acquisition volumes can be considered, as follows:

- Brain and head and neck: In case of the brain, the volume extends from the vertex (upper limit) to the occipital foramen (lower limit). In the head and neck district, the volume extends from a tangential plane to the upper edge of the sellar back (upper limit) up to a floor 2-cm caudal to the upper contour of the sternal manubrium (lower limit).
- Mediastinum: The volume of CT acquisition extends from the cricoid cartilage (upper limit) up to the second lumbar vertebra (lower limit) [12].
- Superior abdomen: The acquisition volume extends from a plane located 2 cm above the dome of the liver (upper limit) to the level of the iliac crests (lower limit).
- Pelvis: The upper limit is established 1 cm cranial to the upper limit of the iliac crests and lower limit at the level of the ischiorectal fossa and – in the presence of rectal tumors infiltrating the anal canal and tumors of the anal canal, vulva, and vaginal canal – at the level of the anal verge.

The main parameters for performing planning CT for these different anatomical regions are listed in Table 4.1. The recommended contrast windows for optimal visualization of the different tissues of the four anatomical regions are the following:

- Brain and head and neck: For the brain, the size of the window and the level of CT images are, respectively, 40 and 80–120 HU. For the head and neck region, the width and window level of CT images for soft tissue are, respectively, 350 and 35 HU. The corresponding settings for the analysis of bone structures are 2,000 HU and 400 HU.
- Mediastinum: An appropriate window to the

Table 4.1 Parameters for performing planning computed tomography (CT)

	Brain	Head and neck	Mediastinum[a]	Upper abdominal region	Pelvis[b]
Slice thickness	4 mm	3 mm	5 mm	5 mm	8 mm
Table speed/ rotation	1.5mm/s	3 mm/s	5-8 mm/s	8 mm/s	10 mm/s
Pitch	1	1	1.0-1.6	1.6	1.25
Reconstruction interval	2 mm	3 mm	5 mm	5 mm	5 mm
kV	120	120-130	120	140	140
mA	180-200	220-240	240	240	240
Algorithm	Soft/standard	Soft/standard (kernel 4-6)	Soft/standard	Soft/standard	Soft/standard
Matrix	512 x 512 pixel	512 x 512 pixel	1,024 x 1,024 pixel	1,024 x 1,024 pixel	1,024 x 1,024 pixel
FOV	250 Variable on the basis of the size of the skull	Adapted to the patient and sized to include the patient's contour	Adapted to the patient and sized to include the patient's contour	Adapted to the patient and sized to include the patient's contour	Adapted to the patient and sized to include the patient's contour

kV kilovolts, *mA* milliampere, *FOV* field of view.

[a]For possible optimization of the CT acquisition method, parameters for this region may be changed as follows: slice thickness 3 mm, table speed/rotation 3 mm/s, reconstruction interval 3 mm.

[b]For possible optimization of the CT acquisition method, parameters for this region may be changed as follows: slice thickness 5 mm, table speed/rotation 8 mm/s, reconstruction interval 4 mm.

mediastinum has a width of 400 HU and a level of +40 HU, whereas for the lung parenchyma, the amplitude recommended is 1,600 HU and –600 HU is the level [16, 17]

- Superior abdomen: For the study of the upper abdomen, the contrast window has a recommended level of 40 HU and a width of 350–400 HU [18].
- Pelvis: For imaging soft tissue, it is advisable to use a width of 400 HU and a level of 40 HU and bone parameters already described for the head and neck [18].

A procedure of coregistration and fusion CT-MRI has been used to identify some critical organs that are difficult to evaluate with CT only. In particular, the combination of CT-MRI improves visualization in the brain district considering the hypothalamus, the pituitary gland, the optic chiasm, and the cochlea; in the head and neck and mediastinum district to identify the constrictor muscles of the pharynx, brachial plexus, and heart; and for the pelvic area in relation to the rectum, penile bulb, and urogenital diaphragm.

The peculiarity of identifying these organs by MRI was represented by the detection of anatomoradiological limits (cranial, caudal, lateral, medial, anterior, posterior) that were easily recognizable on conventional planning CT scans.

Acknowledgement This chapter has been written with the contributions of Raffaella Basilico, Massimo Caulo, Antonella Filippone, and Rossella Patea.

References

1. Levitt SH, Purdy JA, Perez CA, Vijayakumar PC (2006) Technical basis of radiation therapy. Practical clinical applications, 4th rev. edn. Springer-Verlag
2. Pelizzari CA (1994) Registration of three-dimensional medical image data. ICRU News 1:4–14
3. Kessler ML, Pitluck S, Petti P, Castro JR (1991) Integration of multimodality imaging data for radiotherapy treatment planning. Int J Radiat Oncol Biol Phys 21:1653–1667

4. Khoo VS, Joon DL (2006) New developments in MRI for target volume delineation in radiotherapy. Br J Radiol 79:S2–S15
5. Hill DLG, Batchelor PG, Holden M, Hawkes DJ (2001) Medical image registration. Phys Med Biol 46:R1–R45
6. Khan FM, Gerbi BJ (2012) Treatment planning in radiation oncology, 3rd edn. Wolters Kluwer/Lippincott William & Wilkins, Philadelphia
7. Ashamalla H, Addeo D, Ikoro NC et al (2003) Commissioning and clinical results utilizing the Gildenbergy-Laitinen adapter device for X-ray in fractionated stereotactic radiotherapy. Int J Radiat Oncol Biol Phys 56:592–598
8. Bova FJ, Meeks SL, Friedman WA et al (1998) Optic-guided stereotactic radiotherapy. Med Dosim 23:221–228
9. Kai J, Shiomi H, Sasama T et al (1998) Optical high-precision three-dimensional position measurement system suitable for head motion tracking in frameless stereotactic radiosurgery. Comput Aided Surg 3:257–263
10. Chang SD, Adler JR (2001) Robotics and radiosurgery: the CyberKnife. Stereotact Funct Neurosurg 76:204–208
11. Yin FF, Zhu J, Yan H et al (2002) Dosimetric characteristics of Novalis shaped beam surgery unit. Med Phys 29:1729–1738
12. Senan S, De Ruysscher D, Giraud P et al, for the Radiotherapy Group, European Organization for Research and Treatment of Cancer (EORTC) (2004) Literature-based recommendations for treatment planning and execution in high-dose radiotherapy for lung cancer. Radiother Oncol 7:139–146
13. Workmanns D, Diederich S, Lentschig MG et al (2000) Spiral CT of pulmonary nodules: interobserver variations in assessment of lesion size. Eur Radiol 10:710–713
14. Armstrong J, Mc Gibney C (2000) The impact of three-dimensional radiation on the treatment of non-small cell lung cancer. Radiother Oncol 56:157–167
15. Lagerwaard FJ, Van Sornsen de Koste JR, Nijssen-Visser MR et al (2001) Multiple slow CT scans for incorporating lung tumor mobility in radiotherapy planning. Int J Radiat Oncol Biol Phys 51: 932–937
16. Harris KM, Adams H, Lloyd DC, Harvey DJ (1993) The effect on apparent size of simulated pulmonary nodules of using three standard CT window settings. Clin Radiol 47:241–244
17. Giraud P (2000) Influence of CT image visualization parameters for target volume delineation in lung cancer. Proceedings of 19th ESTRO Istanbul 2000. Radiother Oncol S39
18. Ausili-Cefaro G, Genovesi D, Perez CA, Vinciguerra A (2008) A guide for delineation of lymph nodal clinical target volume in radiation therapy. Springer-Verlag, New York

Part III

Axial CT Radiological Anatomy: Image Gallery

Anatomical Reference Points

1 – Cranial theca
2 – Clinoid processes
3 – Uncus hyppocampi
4 – Mesencephalon
5 – Cavum sellae
6 – Muscles (rect, medial and lateral)
7 – Medial wall of tympanic cavity
8 – Internal auditory canal
9 – Occipital condyle
10 – Mandibular branch
11 – Masseter muscle
12 – Sternocleidomastoid muscle
13 – Longus capitis muscle
14 – Pterygoid muscle
15 – Palatine tonsil
16 – Hyoid bone
17 – Middle scalene muscle
18 – Thyroid cartilage
19 – Anterior scalene muscle
20 – Arytenoid cartilage
21 – Cricoid cartilage
22 – Tracheal ring
23 – Clavicle
24 – Subclavian artery

Anatomical Boundaries

ORGAN AT RISK	CRANIAL	CAUDAL	LATERAL
Cochlea	Petrous apex of temporal pyramid	Carotid canal	Medial wall of tympanic cavity
Optic chiasm	0.5 cm above anterior clinoid processes	Cavum sellae	Medial surface of uncus hippocampi, temporal horn of lateral ventricle
Brachial plexus	Neural foramina C4-C5	First half of clavicular head	Sternocleidomastoid muscle, subclavian and axillary neurovascular bundle
Superior pharyngeal constrictor muscle	Occipital condyle	Superior edge of hyoid bone	Palatine tonsil or parapharyngeal space
Middle pharyngeal constrictor muscle	0.5 cm above hyoid bone	Inferior edge of the hyoid bone	Palatine tonsil or parapharyngeal space
Inferior pharyngeal constrictor muscle	Inferior edge of the hyoid bone	Esophagus	Parapharyngeal space, lateral edge of thyroid
Parotid	Inferior edge of external auditory canal	Inferior margin of mandibular branch	Subcutaneus adipose tissue
Optic bulb and optic nerve	Superior rectus muscle, adipose tissue	Inferior rectus muscle, adipose tissue	Lateral rectus muscle, adipose tissue
Brain	Internal edge of frontal and parietal bones	Internal edge of occipital and temporal bones	Internal edge of temporal, occipital, and parietal bones
Lens	Hyperdense area within the optic bulb (anterior surface)		

Chapter 5.1
Brain, Head and Neck

Color Legend

- Brain
- Ocular bulb
- Optic chiasm
- Hypothalamic infundibulum
- Adenohypophysis
- Pituitary stalk
- Cristalline
- Cochlea
- Pharingeal constrictor muscles
- Parotid
- Spinal cord
- Brain stem
- Mandible
- Brachial plexus
- Larynx

MEDIAL	ANTERIOR	POSTERIOR	CT window
Temporal pyramid	Anterior and superior surface of petrous bone	Anterior aspect of internal auditory canal	Bone: C450, W1,600
	Line through the anterior clinoid processes	Mesencephalon (anterior aspect of cerebral pedunculi), posterior clinoid included	Brain: C35, W100 Bone: C450, W1,600
C4-T1 neural foramina; C4-T1 vertebral peduncle	Neck vascular bundle (C4-C6), Anterior scalenus muscle (C6-T1)	Middle scalene muscle first rib, subclavian vein	H&N: C35, W350
Pharynx	Pterygoidei muscle: anterior margin	Longus capitis muscle	H&N: C35, W350
Pharynx	Oropharynx, Lateral edge of hyoid bone	Longus capitis muscle	H&N: C35, W350
Pharynx	Arytenoid, first tracheal ring	Longus capitis muscle, anterior vertebral body	H&N: C35, W350
Longus capitis muscle and internal pterygoidei muscle	Posterior border of masseter muscle (clavicular branch), mandibular posterior corner	Sternocleidomastoid muscle	H&N: C35, W350
Medial rectus muscle, Adipose tissue		Retrobulbar adipose tissue, optic foramen	H&N: C35, W350 Bone: C450, W1,600 (for the intracanalar tract of the nerve)
	Internal edge of frontal bone and sphenoid bone	Internal edge of occipital bone	Brain: C35, W100 Bone: C450, W1,600
Hyperdense area within the optic bulb (anterior surface)			

G. Ausili Cefaro, D. Genovesi, C.A. Perez, *Delineating Organs at Risk in Radiation Therapy*,
DOI: 10.1007/978-88-470-5257-4_5-1, © Springer-Verlag Italia 2013

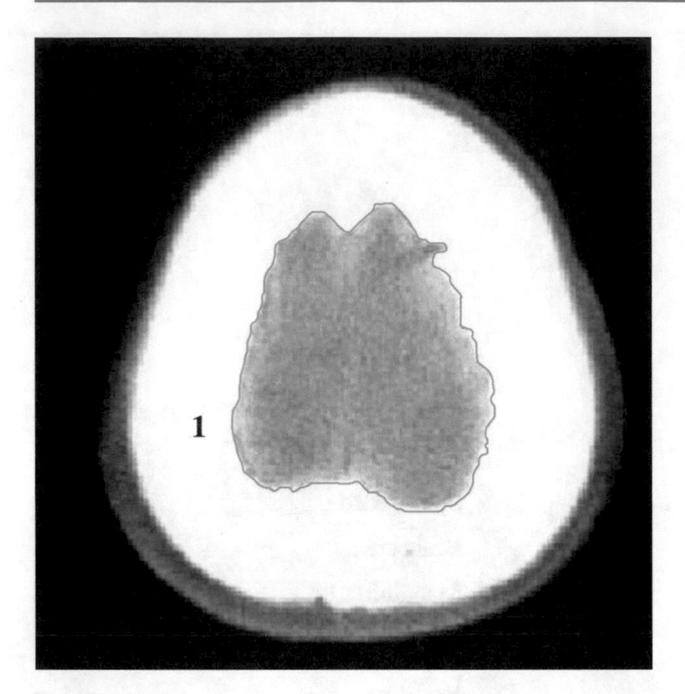

Fig. 5.1

Brain

1 – Cranial theca

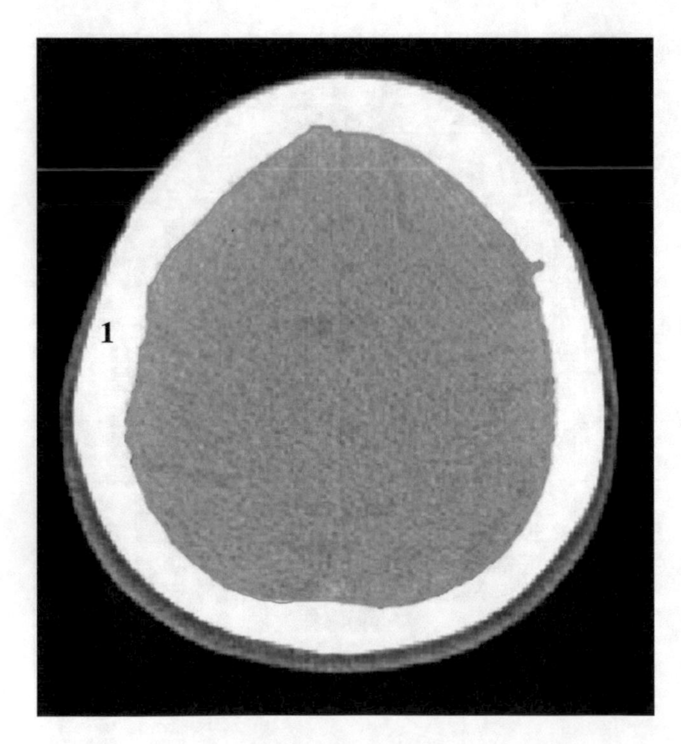

Fig. 5.2

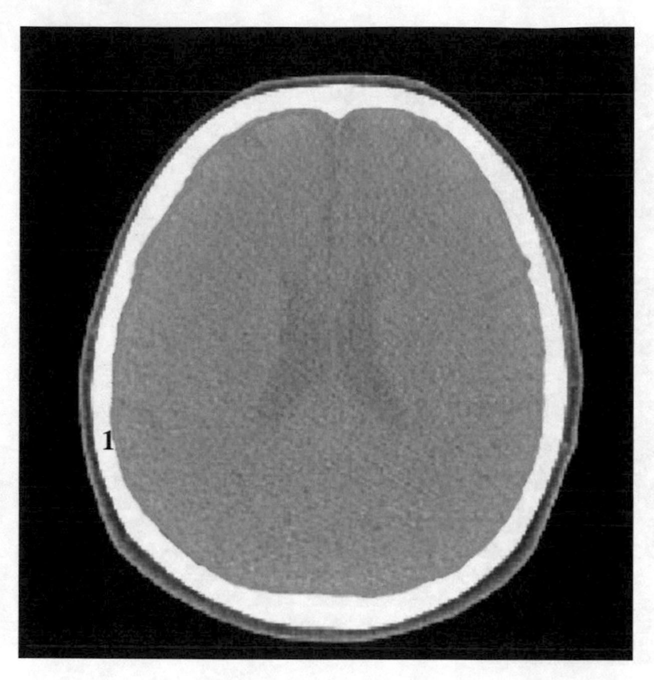

Fig. 5.3

Brain
Ocular bulb

1 – Cranial theca

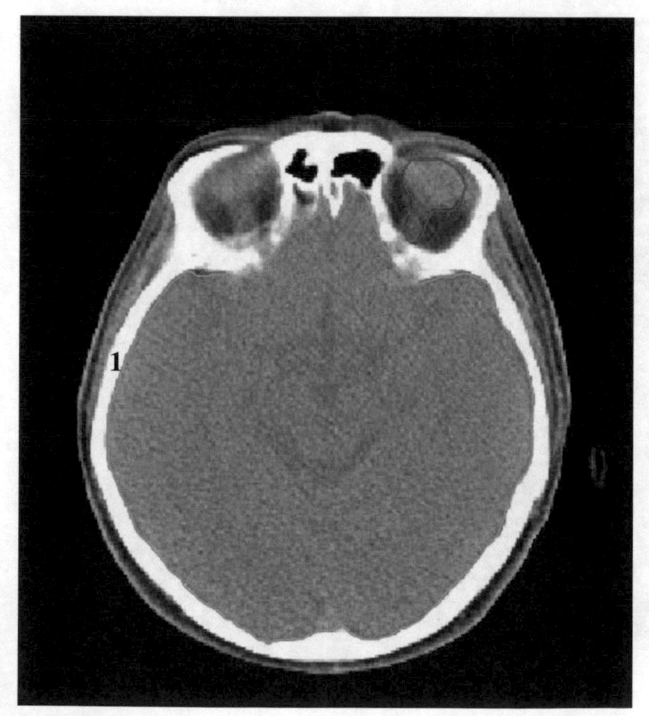

Fig. 5.4

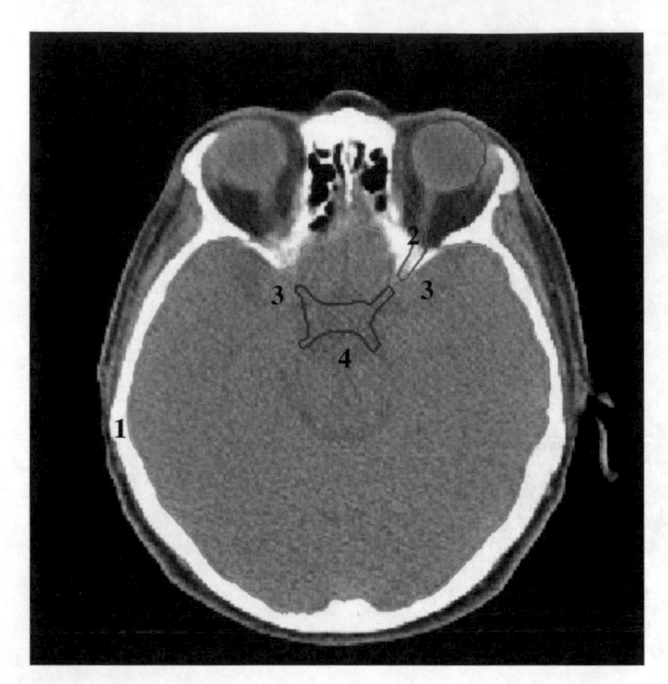

Brain

Ocular bulb

Optic chiasm

1 – Cranial theca

2 – Clinoid processes

3 – Uncus hyppocampi

4 – Mesencephalon

Fig. 5.5a

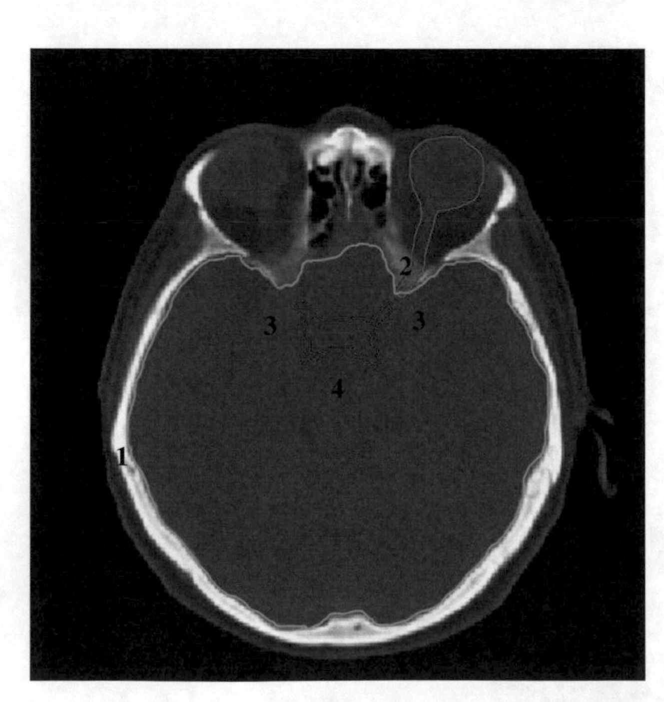

Fig. 5.5b Bone window

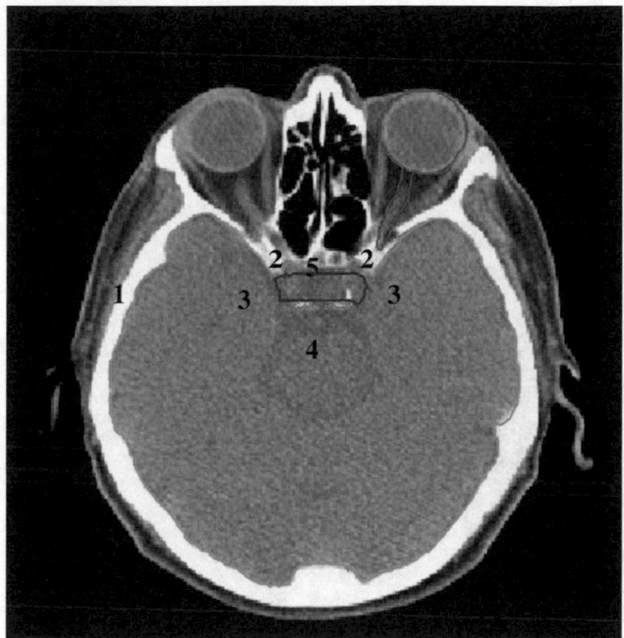

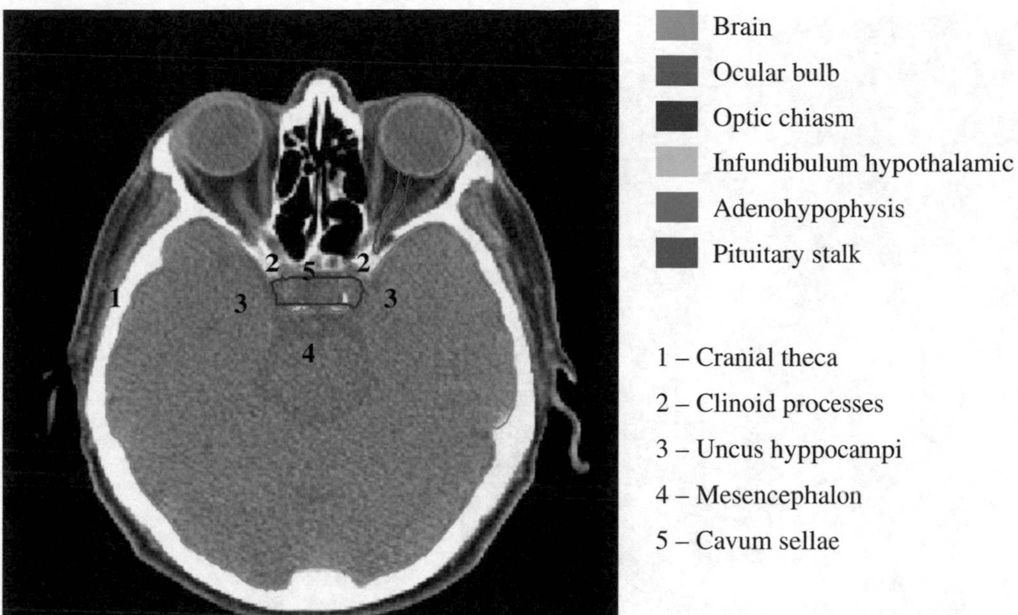

Fig. 5.6a

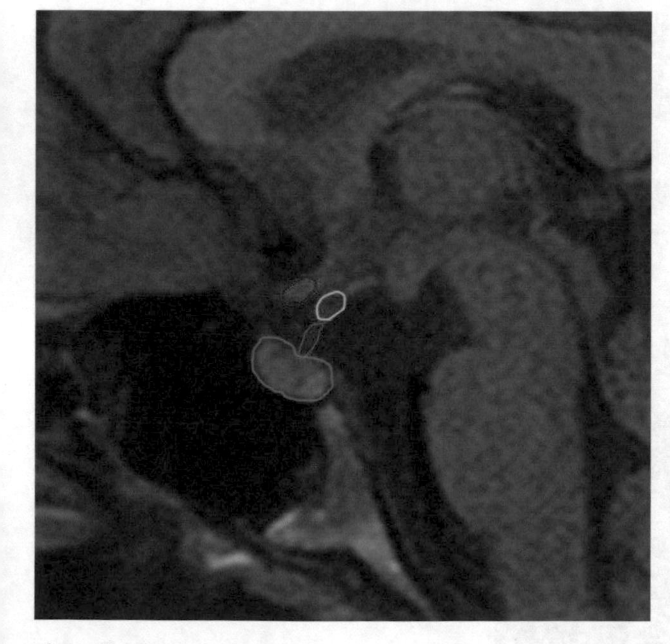

Fig. 5.6b MRI sagittal view

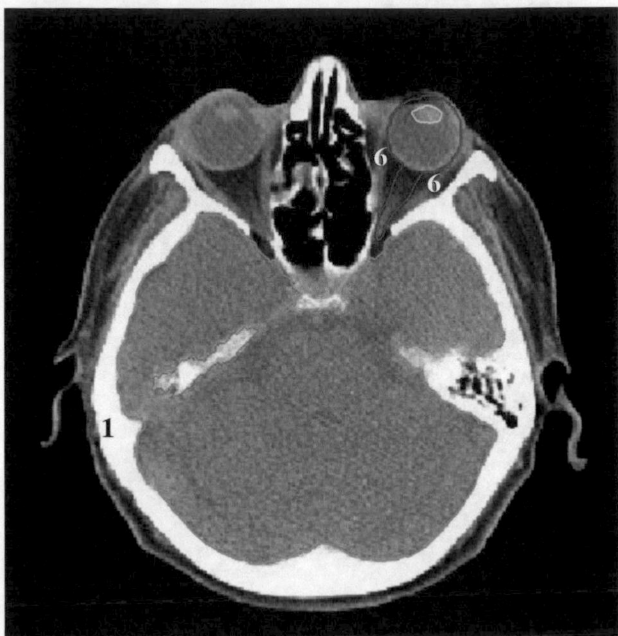
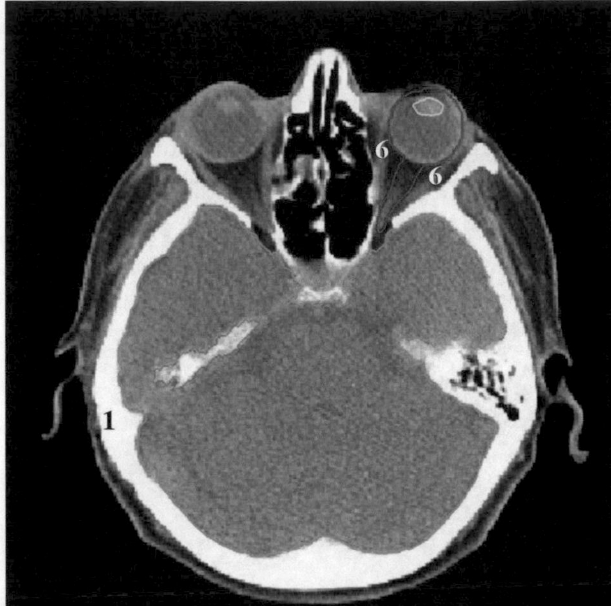

Brain

Ocular bulb

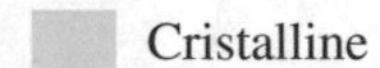
Cristalline

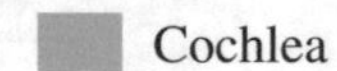
Cochlea

1 – Cranial theca

6 – Muscles (rect, medial and lateral)

7 – Medial wall of tympanic cavity

8 – Internal auditory canal

Fig. 5.7

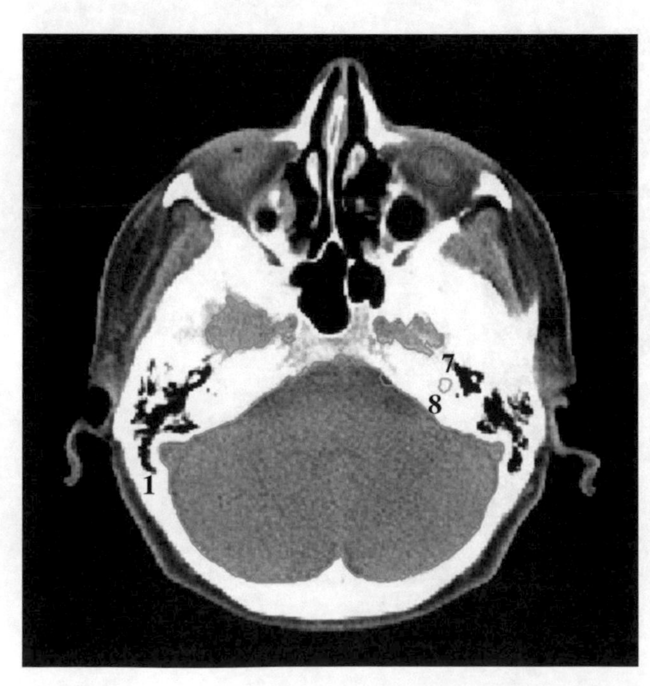

Fig. 5.8

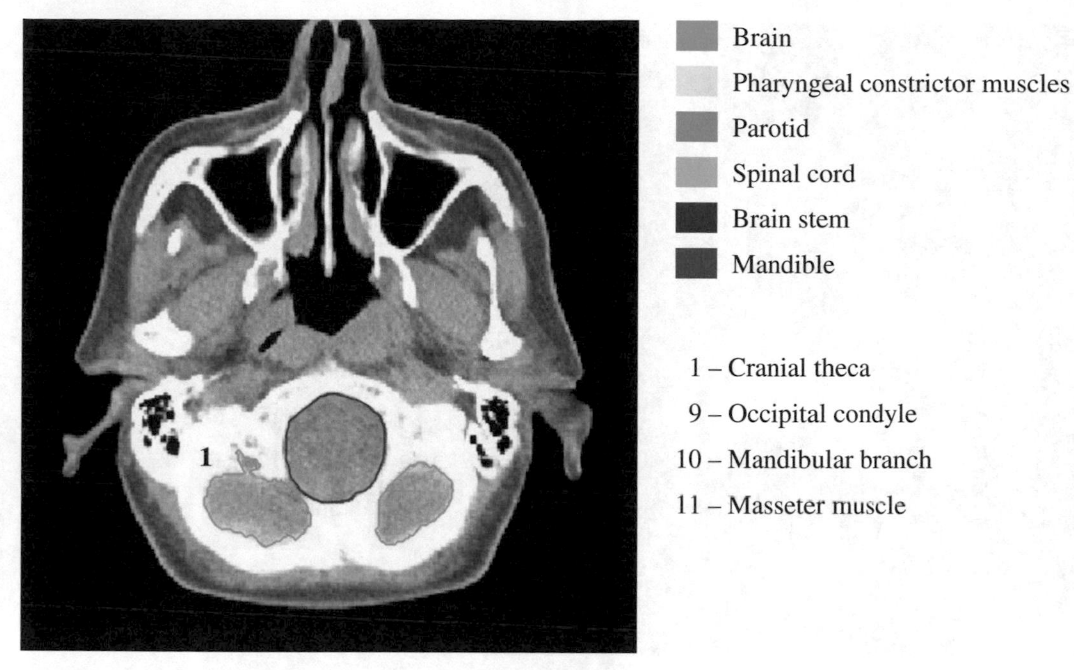

Fig. 5.9

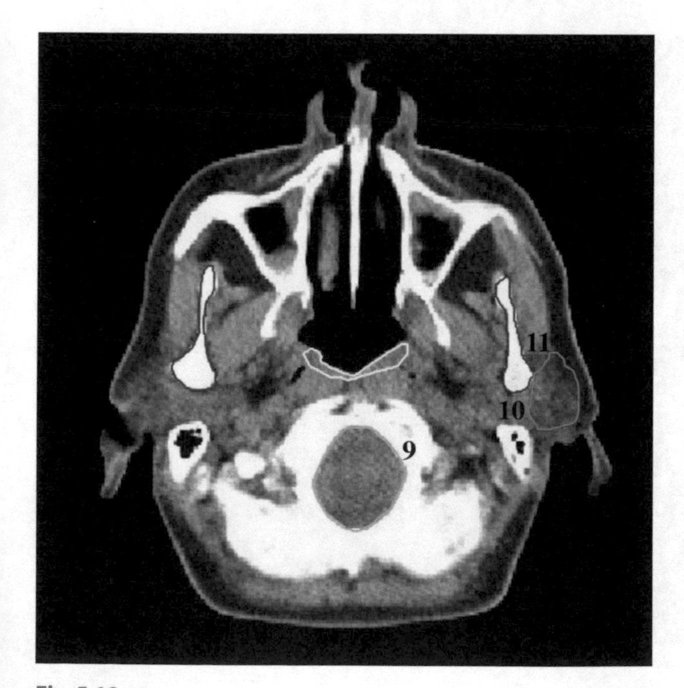

Fig. 5.10

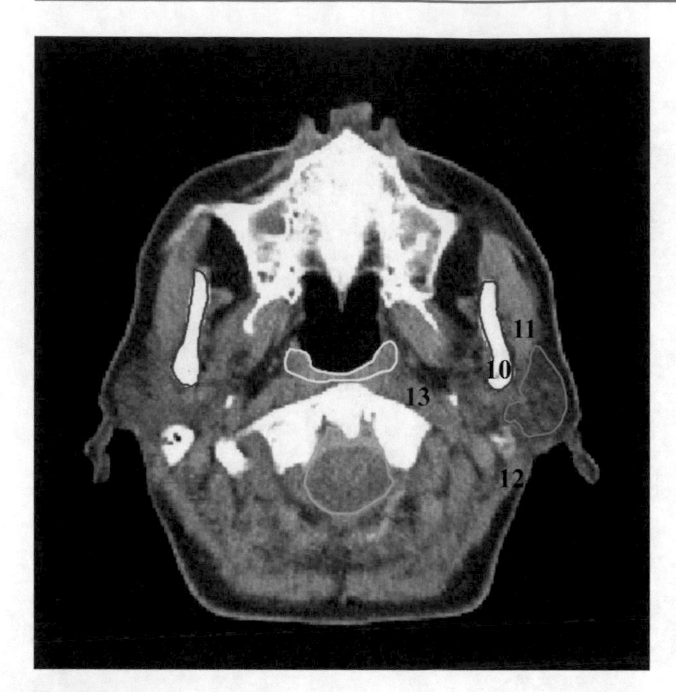

Pharyngeal constrictor muscles
Parotid
Spinal cord
Mandible

10 – Mandibular branch
11 – Masseter muscle
12 – Sternocleidomastoid muscle
13 – Longus capitis muscle
14 – Pterygoid muscle
15 – Palatine tonsil

Fig. 5.11

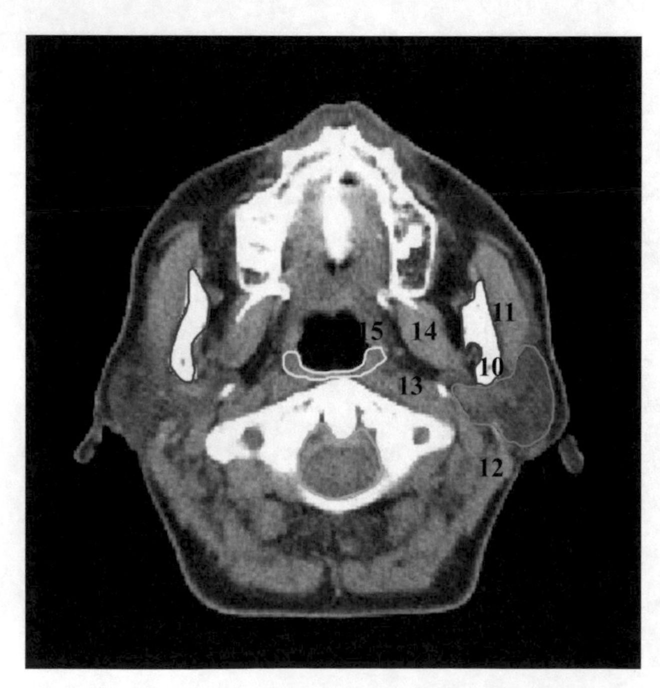

Fig. 5.12

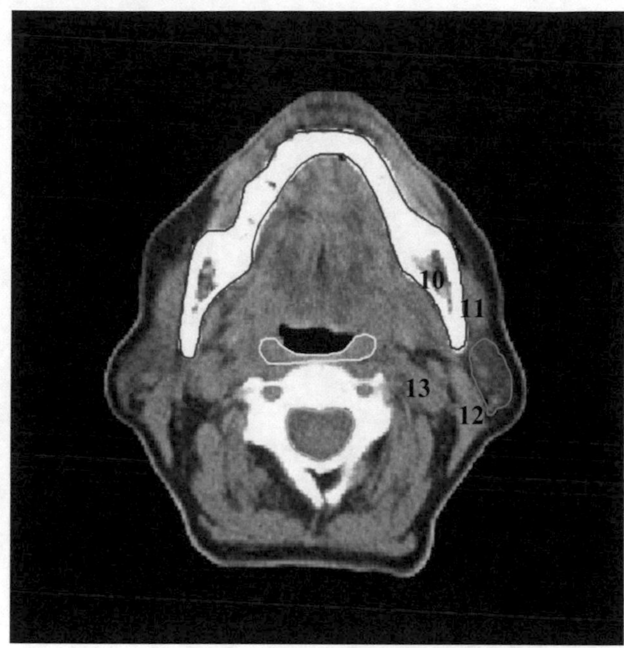

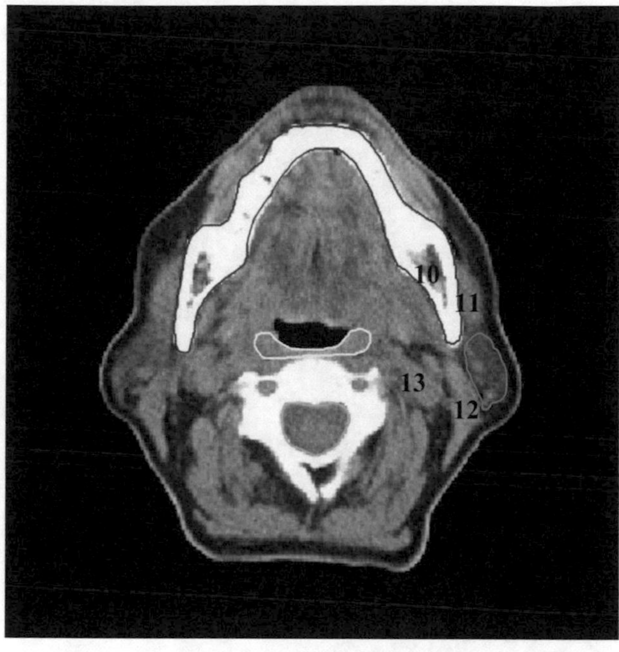

Pharyngeal constrictor muscles
Parotid
Spinal cord
Mandible

10 – Mandibular branch
11 – Masseter muscle
12 – Sternocleidomastoid muscle
13 – Longus capitis muscle

Fig. 5.13

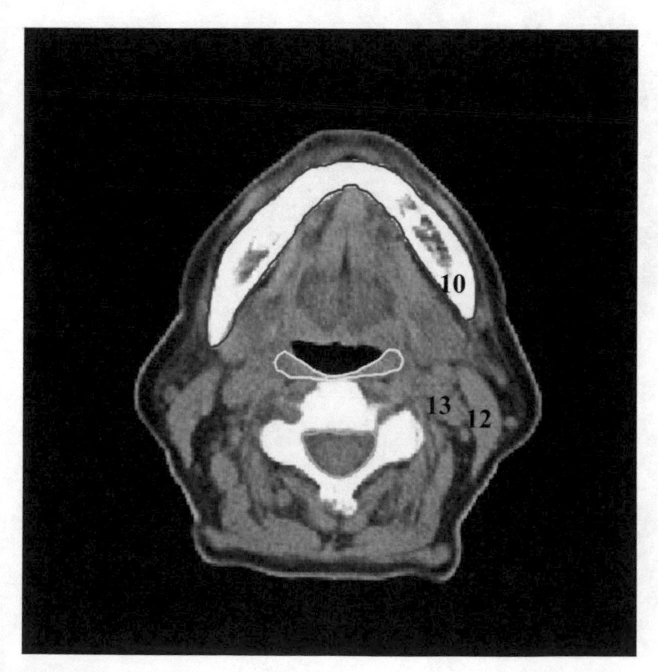

Fig. 5.14

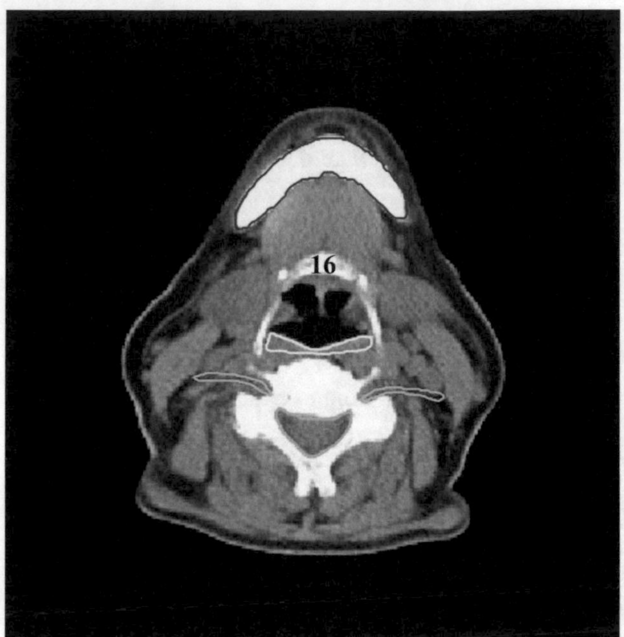

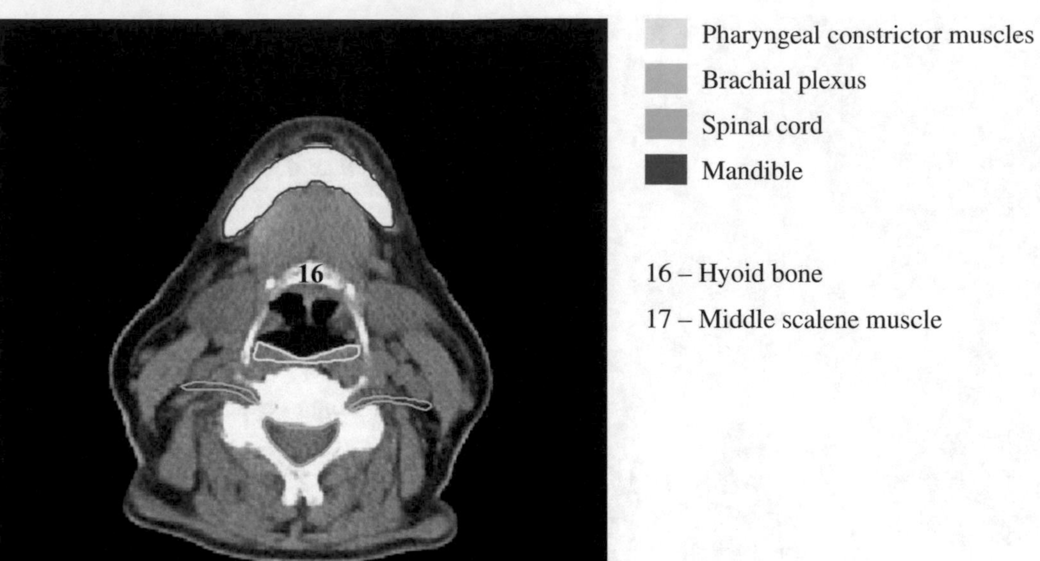

16 – Hyoid bone

17 – Middle scalene muscle

Fig. 5.15

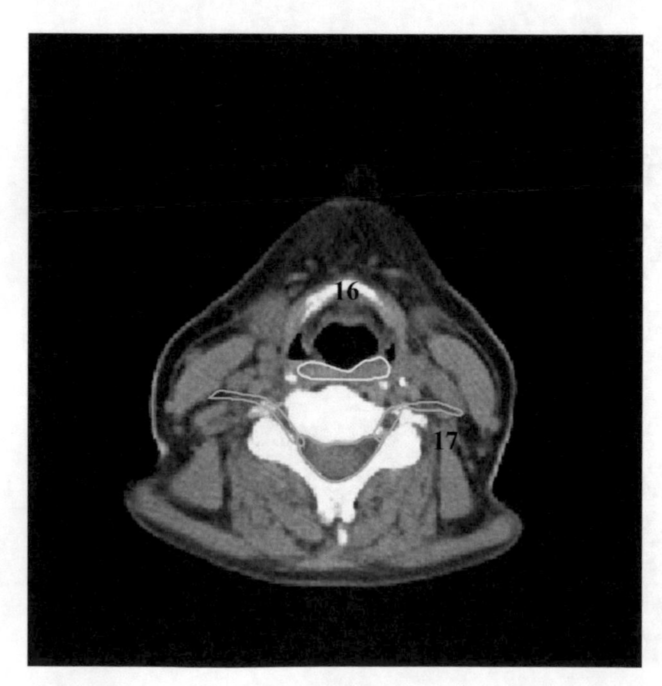

Fig. 5.16

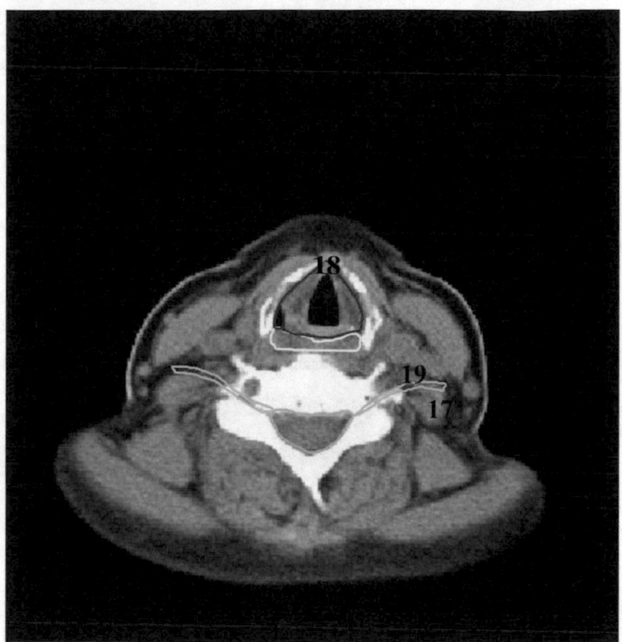
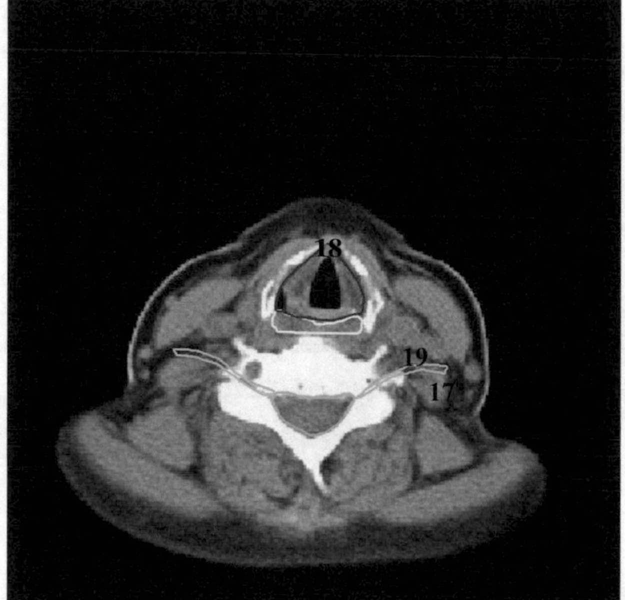

Fig. 5.17

Pharyngeal constrictor muscles
Brachial plexus
Spinal cord
Larynx

17 – Middle scalene muscle
18 – Thyroid cartilage
19 – Anterior scalene muscle
20 – Arytenoid cartilage

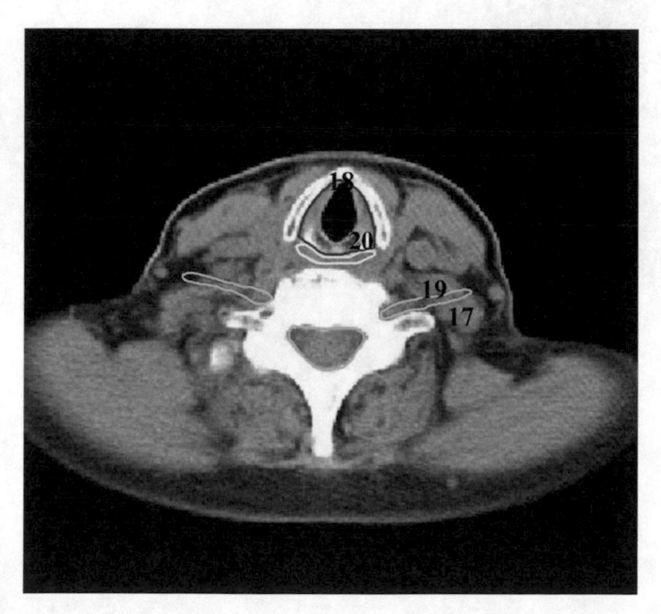

Fig. 5.18

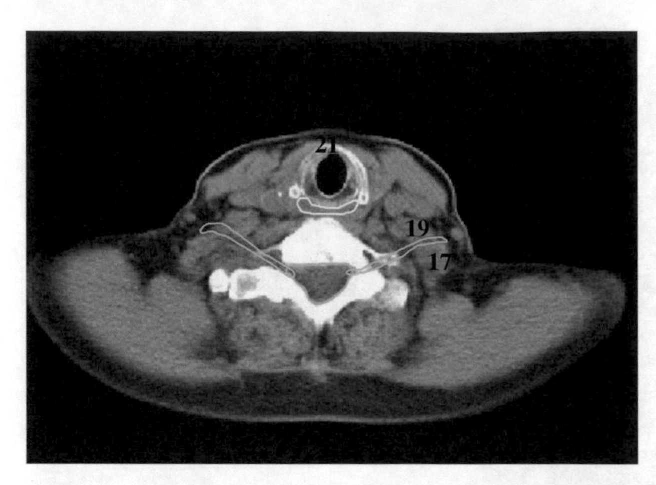

Fig. 5.19

Pharyngeal constrictor muscles

Brachial plexus

Spinal cord

17 – Middle scalene muscle

19 – Anterior scalene muscle

21 – Cricoid cartilage

22 – Tracheal ring

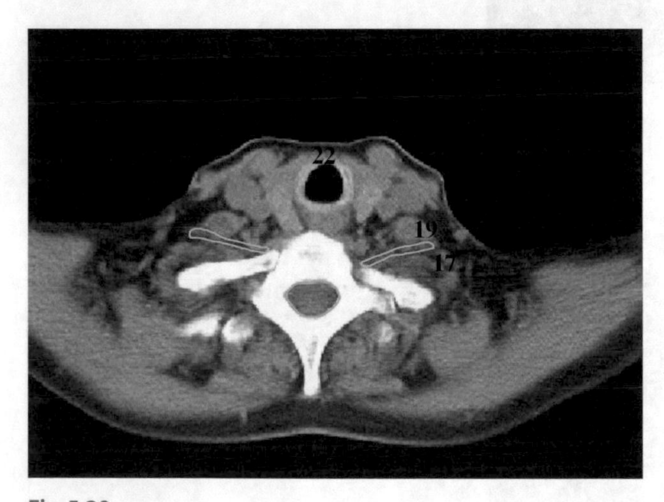

Fig. 5.20

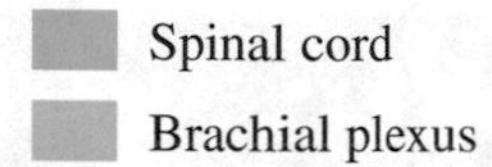

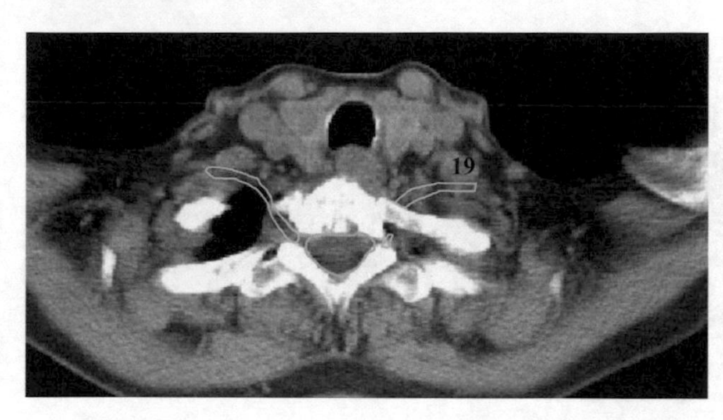

Fig. 5.21

19 – Anterior scalene muscle
23 – Clavicle

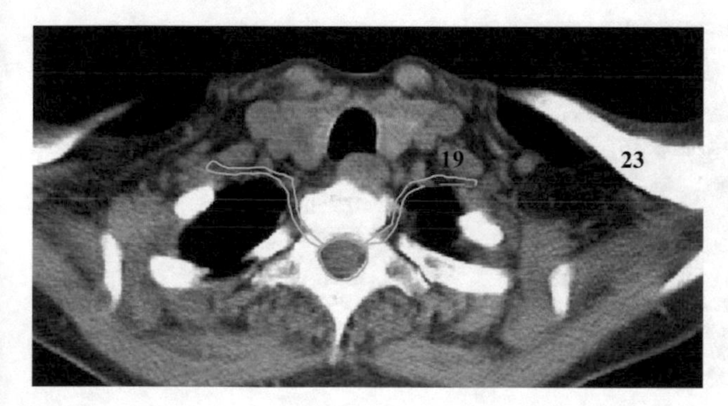

Fig. 5.22

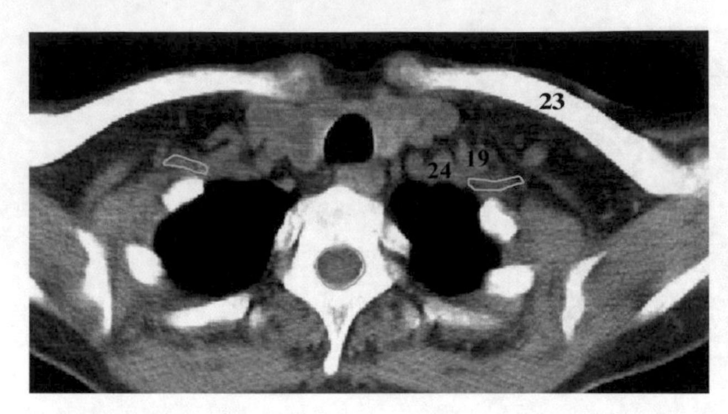

Fig. 5.23

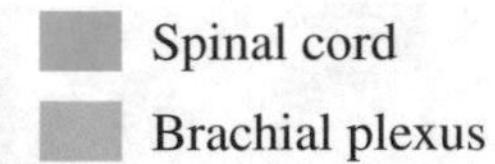

19 – Anterior scalene muscle

23 – Clavicle

24 – Subclavian artery

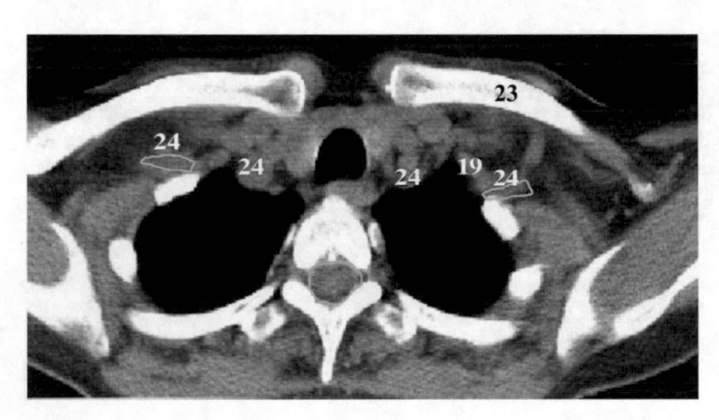

Fig. 5.24

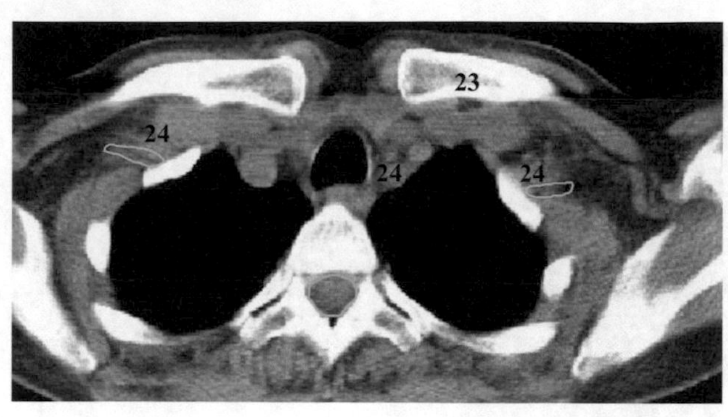

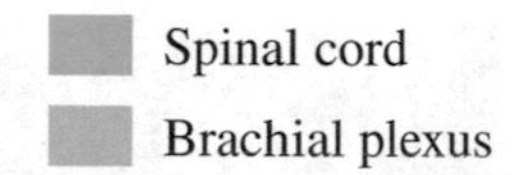

23 – Clavicle

24 – Subclavian artery

Fig. 5.25

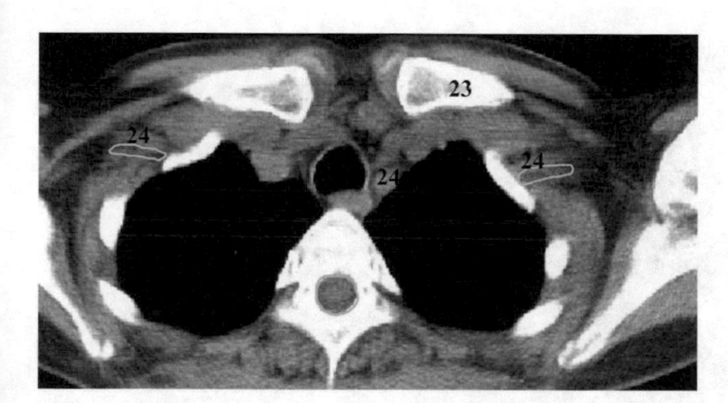

Fig. 5.26

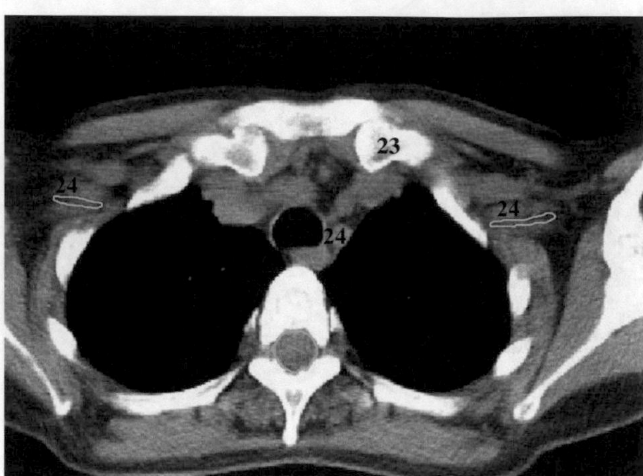

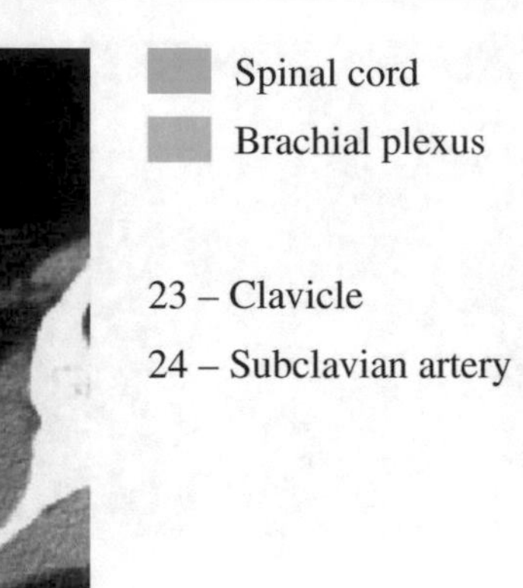

23 – Clavicle

24 – Subclavian artery

Fig. 5.27

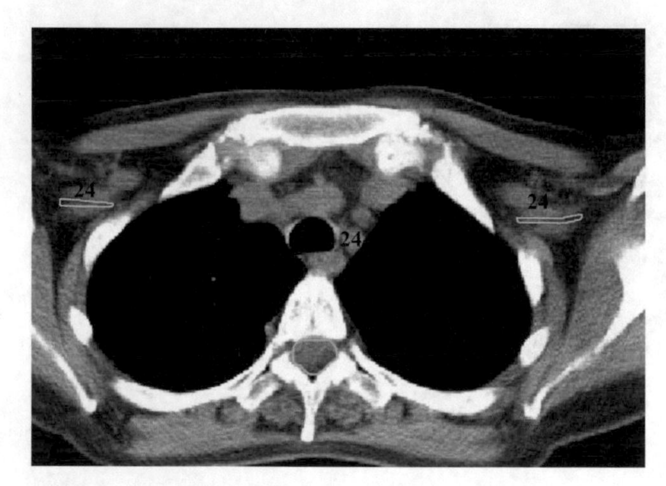

Fig. 5.28

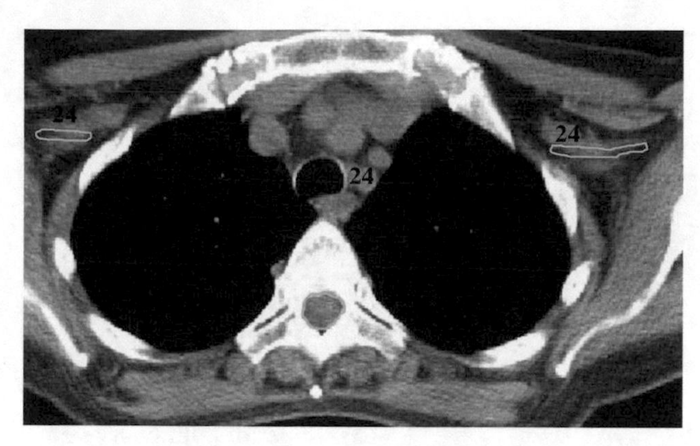

Fig. 5.29

Spinal cord
Brachial plexus

24 – Subclavian artery

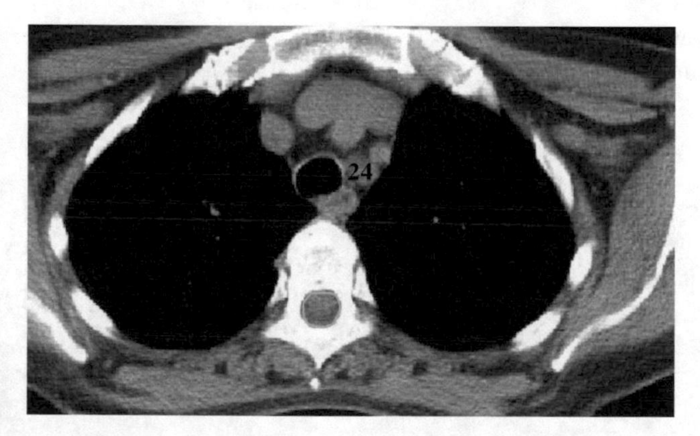

Fig. 5.30

Acknowledgement This chapter has been written with the contributions of Angelo Di Pilla, Massimo Caulo, Annamaria Vinciguerra, Marianna Trignani, and Monica Di Tommaso.

Anatomical Reference Points

1 – Thyroid cartilage
2 – Cricoid cartilage
3 – Deltoid muscle
4 – Infraspinatus muscle
5 – Supraspinatus muscle
6 – Right subclavian artery
7 – Left common carotid artery
8 – Rib
9 – Tracheal ring
10 – Subclavian artery
11 – Aortic arch
12 – Superior vena cava
13 – Aorta
14 – Descending aorta
15 – Pulmonary artery
16 – Superior pulmonary vein
17 – Arch of azygos vein
18 – Azygos vein
19 – Inferior pulmonary vein
20 – Liver
21 – Stomach

Anatomical Boundaries

ORGAN AT RISK	CRANIAL	CAUDAL	LATERAL
Humerus head	Line throught the humeral anatomical neck	Last scan which shows the head	Deltoid muscle,
Mainstem bronchus	Line through the carina	Line through the inferior pulmonary vein	- Aortic arch Left: - Left principal pulmonary artery - Descending aorta Right: - Arch of azygos vein - Superior pulmonary vein - Vena cava
Lung	Line through posterior arch of first rib	Diaphragm	Chest wall
Heart and pericardium	Line through the inferior aspect of left pulmonary trunk	Line through the superior aspect of left hepatic lobe	Mediastinal pleura and lung parenchyma
Esophagus	Caudal aspect of cricoid cartilage	Esophagogastric junction	Left: - Left subclavian artery - Aortic arch - Descendin aorta Right: - Mediastinal pleura and lung parenchyma - Azygos vein - Lung parenchyma
Spinal cord	Occipital condyle	Inferior surface of L2	Vertebral canal
Brachial plexus	Neural foramina C4-C5	First half of clavicular head	Sternocleidomastoid muscle, Subclavian and axillary neurovascular bundle

Chapter 5.2
Mediastinum

Color Legend

- Spinal cord
- Esophagus
- Humerus head
- Left lung
- Main stem bronchus
- Heart and pericardium
- Right atrium
- Anterior interventricular branch of left coronary artery
- Left ventricle
- Right atrium
- Circumflex left artery
- Right ventricle
- Right coronary artery

MEDIAL	ANTERIOR	POSTERIOR
Infraspinatus muscle	Deltoid muscle, Subscapsularis muscle	Supraspinatus muscle
	- Ascending aorta - Left pulmonary artery - Left atrium - Lung parenchyma	- Esophagus - Azygos vein Left: - Descending aorta - Pulmonary vein - Lung parenchyma
Large mediastinal vessels and heart chambers	Chest wall	Costovertebral wall
	Adipose tissue of anterior mediastinum	Esophagus and descending aorta
	- Pars membranacea trachea - Carina - Left main stem bronchus - Posterior wall of left atrium - Adipose tissue of thoraco-abdominal junction	- Vertebral body - Azygos and arch of azygos vein
	Vertebral canal	Vertebral canal
C4-T1 neural foramina; C4-T1 vertebral peduncle	Neck vascular bundle (C4-C6), Anterior scalene muscle (C6-T1)	Middle scalene muscle, First rib, Subclavian vein

G. Ausili Cefaro, D. Genovesi, C.A. Perez, *Delineating Organs at Risk in Radiation Therapy*,
DOI: 10.1007/978-88-470-5257-4_5-2, © Springer-Verlag Italia 2013

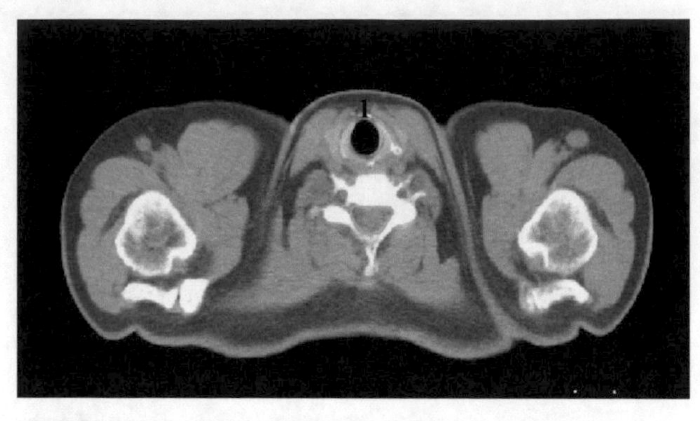

Fig. 5.31

Spinal cord

Esophagus

Head of left humerus

1 – Thyroid cartilage

2 – Cricoid cartilage

3 – Deltoid muscle

4 – Infraspinatus muscle

5 – Supraspinatus muscle

6 – Right subclavian artery

7 – Left common carotid artery

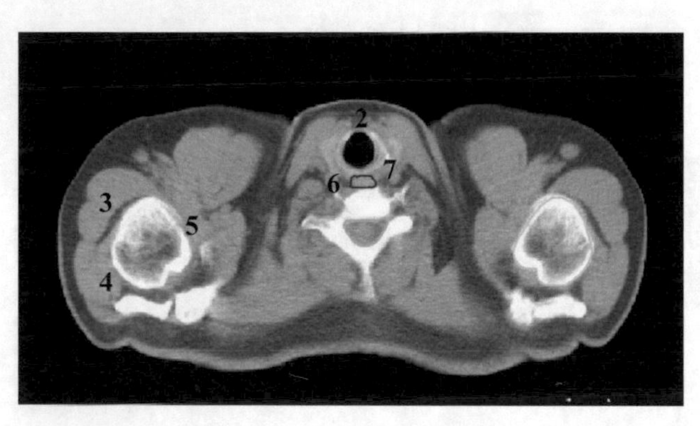

Fig. 5.32

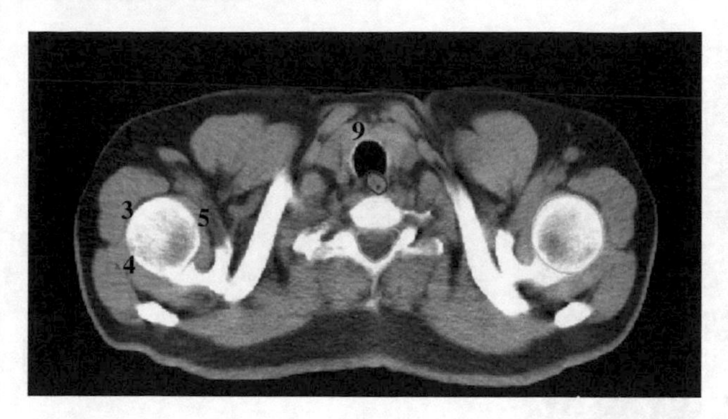

Fig. 5.33

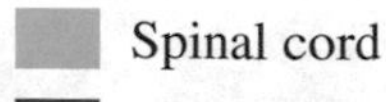

Spinal cord

Esophagus

Head of left humerus

3 – Deltoid muscle

4 – Infraspinatus muscle

5 – Supraspinatus muscle

8 – Rib

9 – Tracheal ring

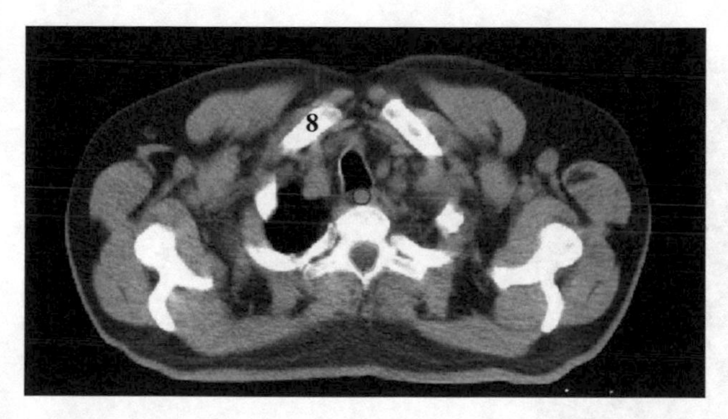

Fig. 5.34

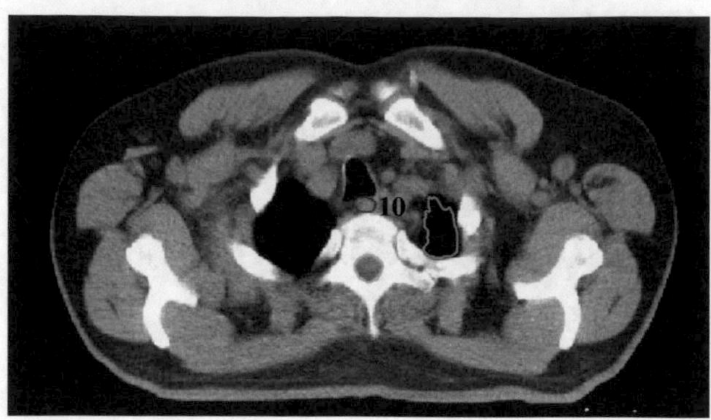

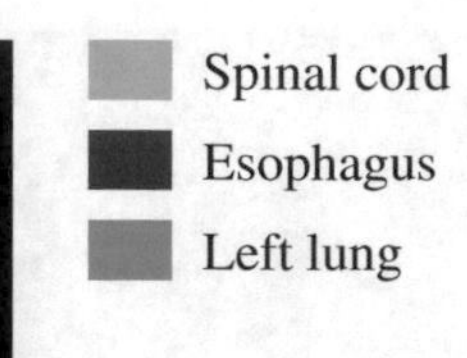

10 – Subclavian artery

Fig. 5.35

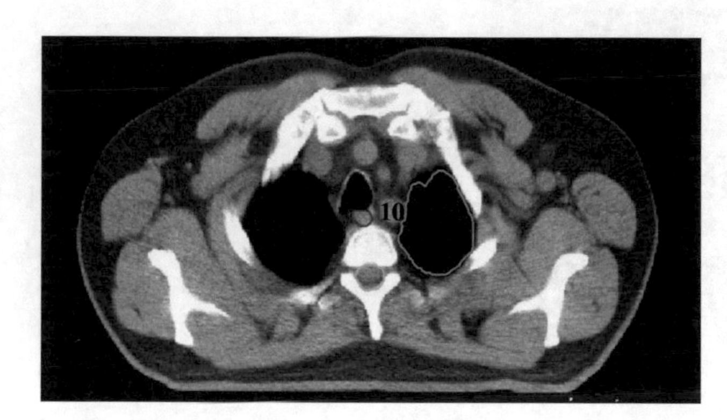

Fig. 5.36

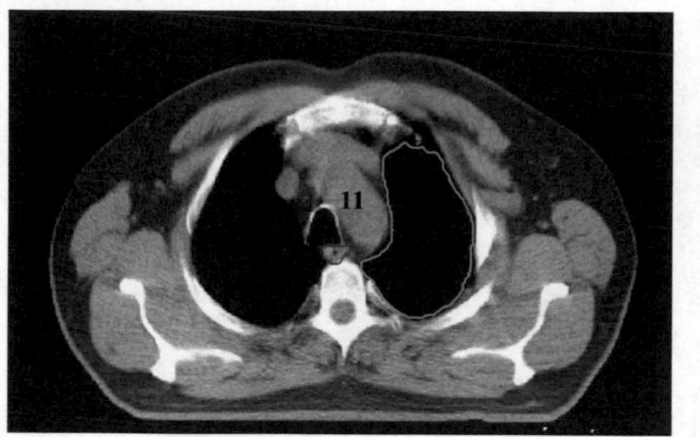

Fig. 5.37

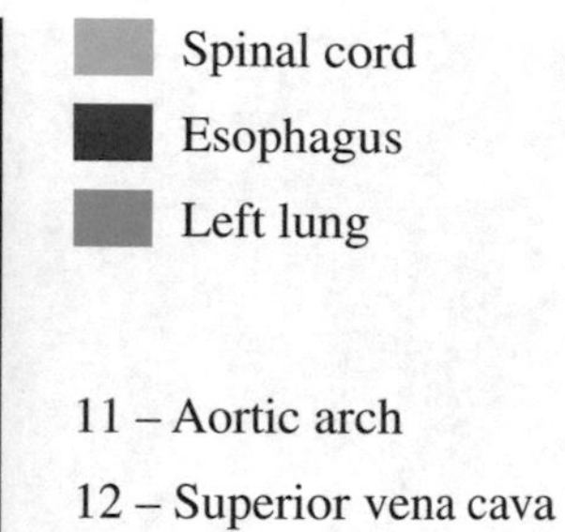

11 – Aortic arch

12 – Superior vena cava

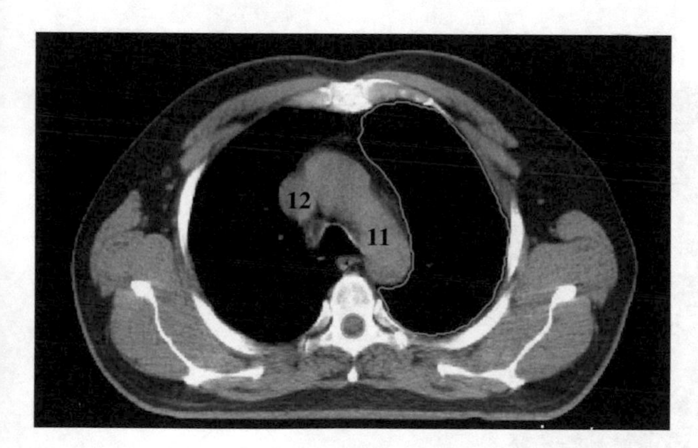

Fig. 5.38

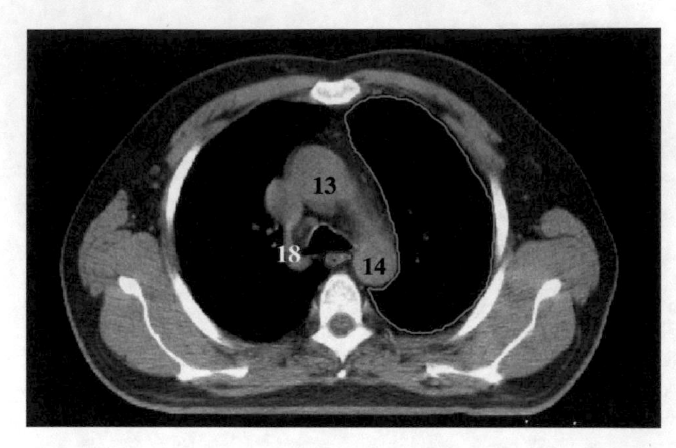

Fig. 5.39

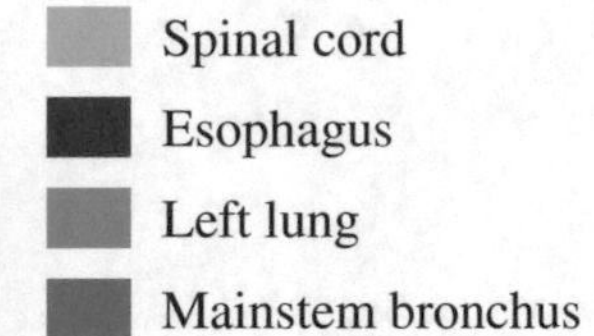

12 – Superior vena cava
13 – Aorta
14 – Descending aorta
15 – Pulmonary artery
16 – Superior pulmonary vein
17 – Arch of azygos vein
18 – Azygos vein

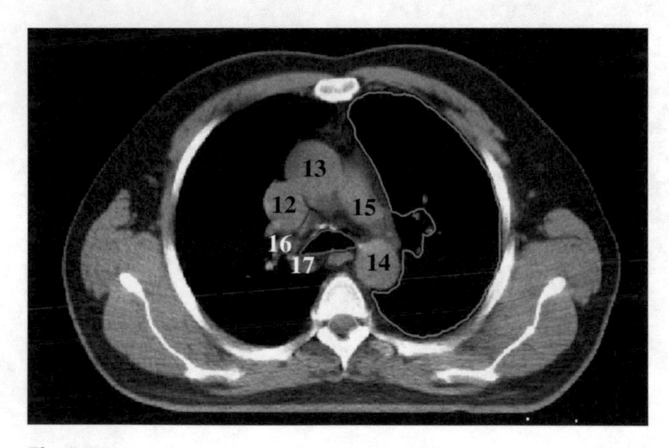

Fig. 5.40

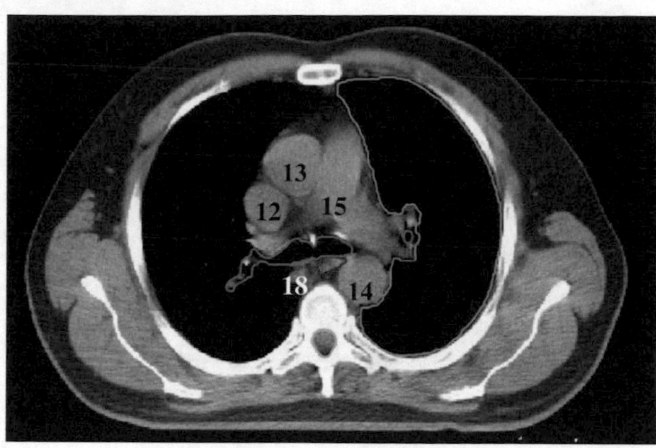

Fig. 5.41

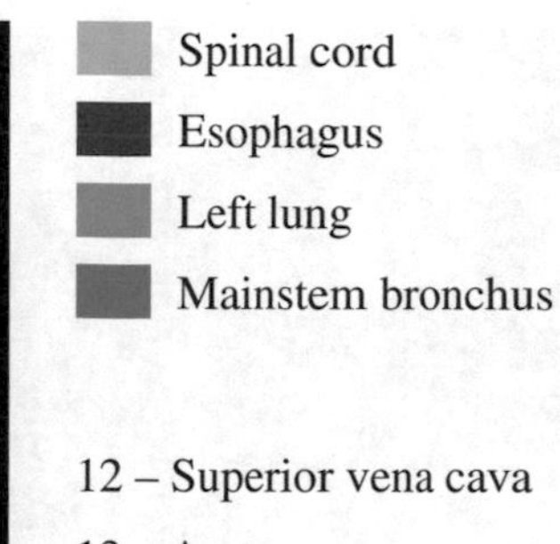

12 – Superior vena cava

13 – Aorta

14 – Descending aorta

15 – Pulmonary artery

18 – Azygos vein

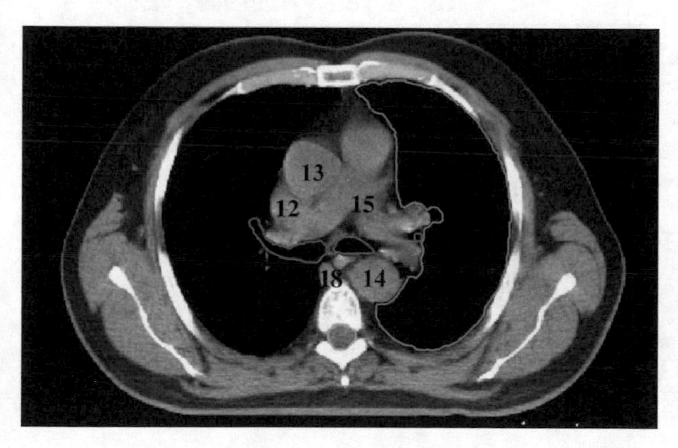

Fig. 5.42

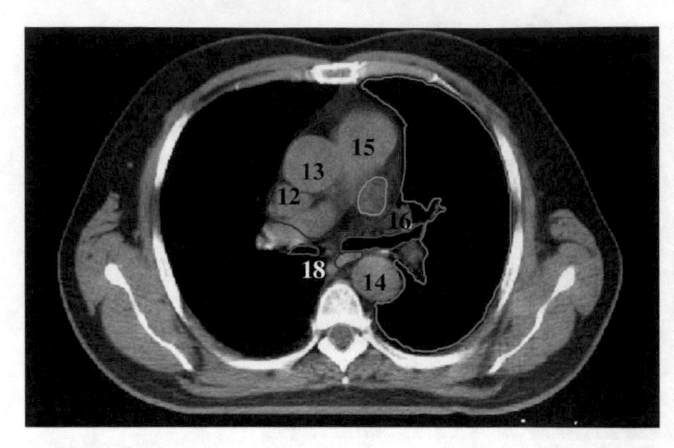

Fig. 5.43

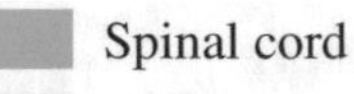

Spinal cord

Esophagus

Left lung

Mainstem bronchus

Heart and pericardium

Right atrium

Anterior interventricular branch of left coronary artery

12 – Superior vena cava

13 – Aorta

14 – Descending aorta

15 – Pulmonary artery

16 – Superior pulmonary vein

18 – Azygos vein

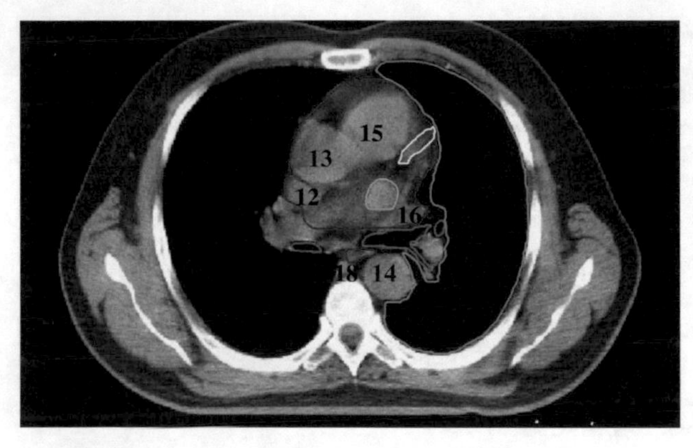

Fig. 5.44

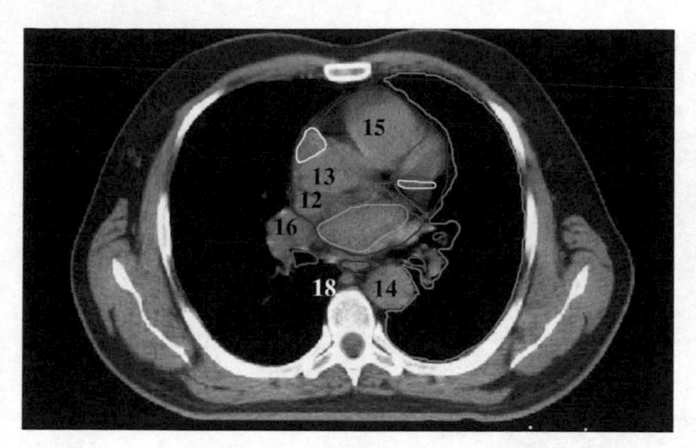

Fig. 5.45

Spinal cord
Esophagus
Left lung
Mainstem bronchus
Heart and pericardium
Right atrium
Anterior interventricular branch of left coronary artery
Left ventricle
Right atrium
Circumflex left artery
Right ventricle

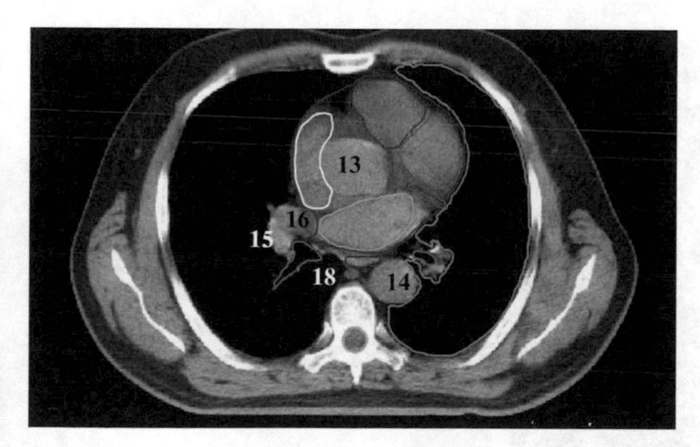

Fig. 5.46

12 – Superior vena cava
13 – Aorta
14 – Descending aorta
15 – Pulmonary artery
16 – Superior pulmonary vein
18 – Azygos vein

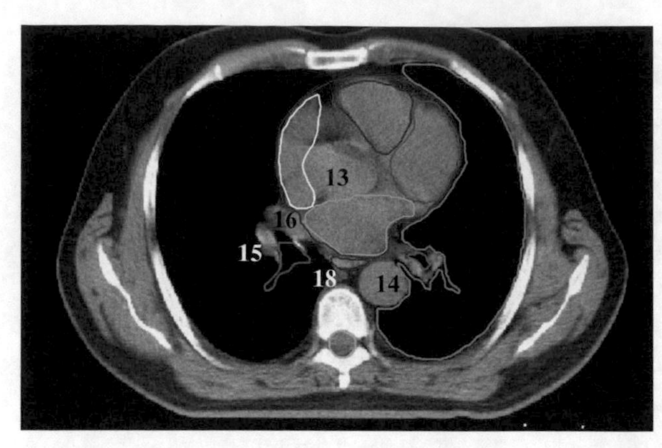

Fig. 5.47

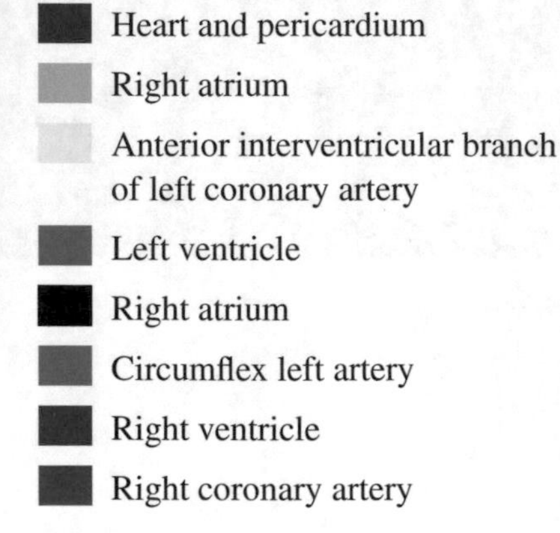

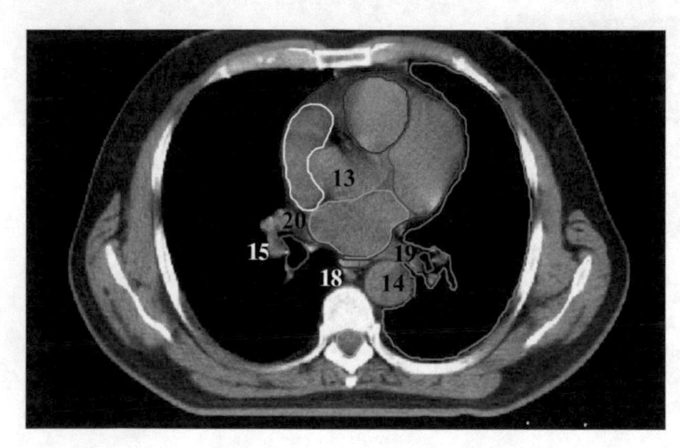

Fig. 5.48

13 – Aorta

14 – Descending aorta

15 – Pulmonary artery

16 – Superior pulmonary vein

18 – Azygos vein

19 – Inferior pulmonary vein

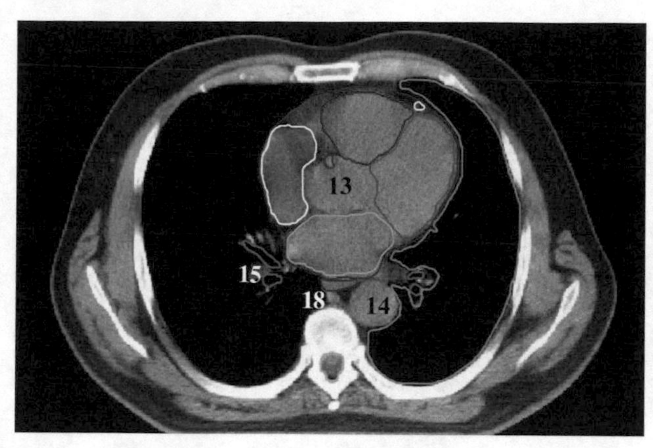

Fig. 5.49

- Spinal cord
- Esophagus
- Left lung
- Mainstem bronchus
- Heart and pericardium
- Right atrium
- Anterior interventricular branch of left coronary artery
- Left ventricle
- Right atrium
- Right ventricle
- Right coronary artery

13 – Aorta
14 – Descending aorta
15 – Pulmonary artery
18 – Azygos vein
19 – Inferior pulmonary vein

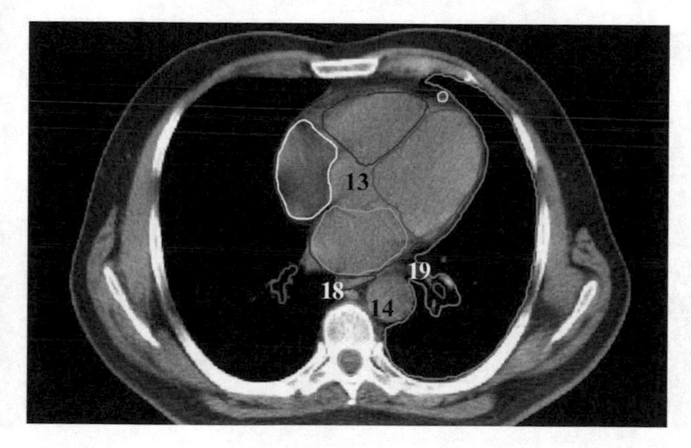

Fig. 5.50

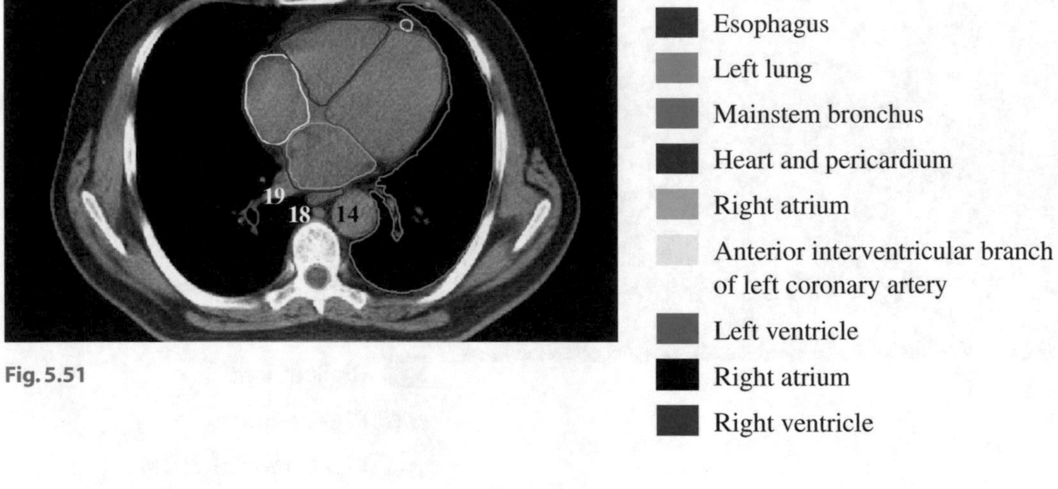

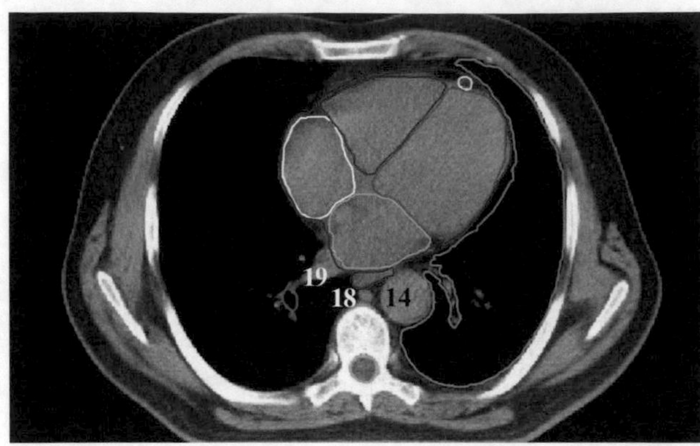

Fig. 5.51

- Spinal cord
- Esophagus
- Left lung
- Mainstem bronchus
- Heart and pericardium
- Right atrium
- Anterior interventricular branch of left coronary artery
- Left ventricle
- Right atrium
- Right ventricle

14 – Descending aorta

18 – Azygos vein

19 – Inferior pulmonary vein

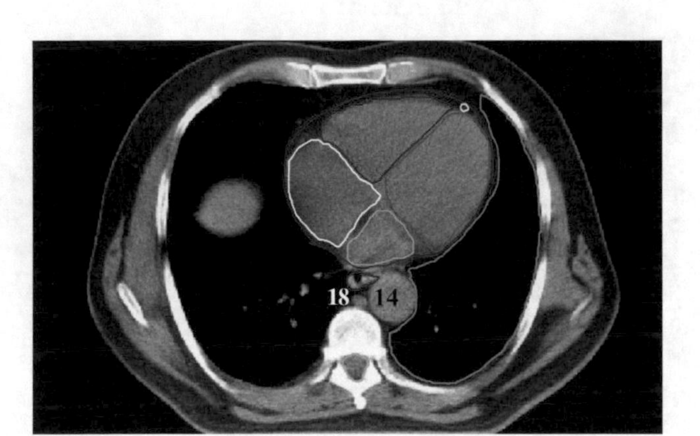

Fig. 5.52

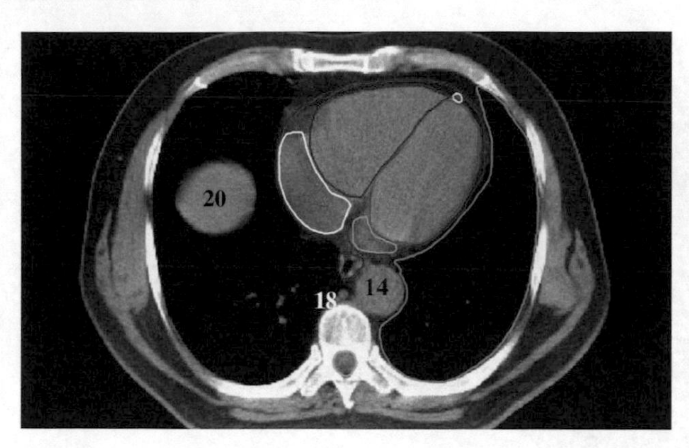

Fig. 5.53

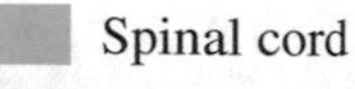

- Spinal cord
- Esophagus
- Left lung
- Mainstem bronchus
- Heart and pericardium
- Right atrium
- Anterior interventricular branch of left coronary artery
- Left ventricle
- Right atrium
- Right ventricle

14 – Descending aorta

18 – Azygos vein

20 – Liver

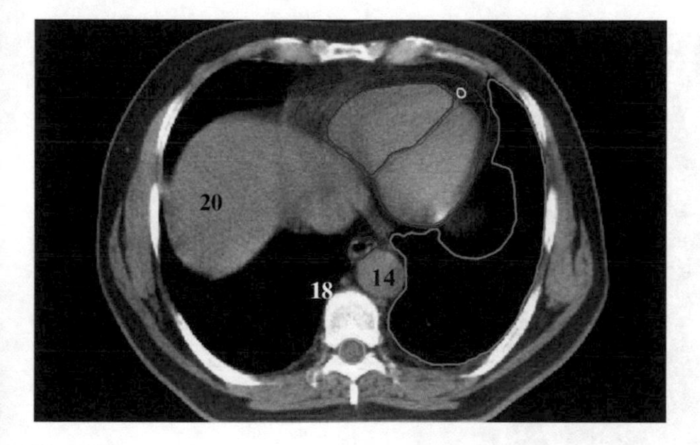

Fig. 5.54

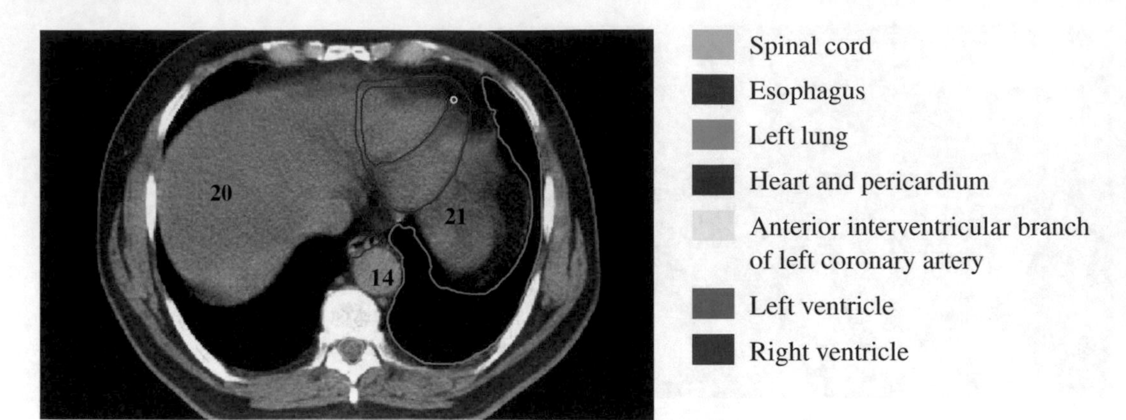

Fig. 5.55

14 – Descending aorta

20 – Liver

21 – Stomach

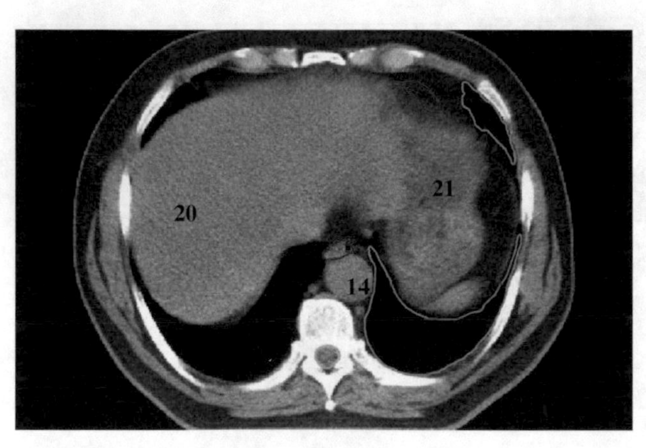

Fig. 5.56

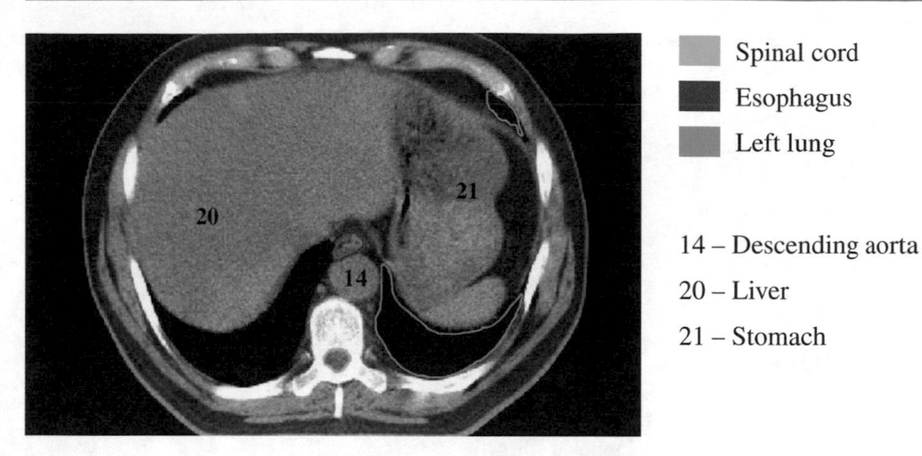

Fig. 5.57

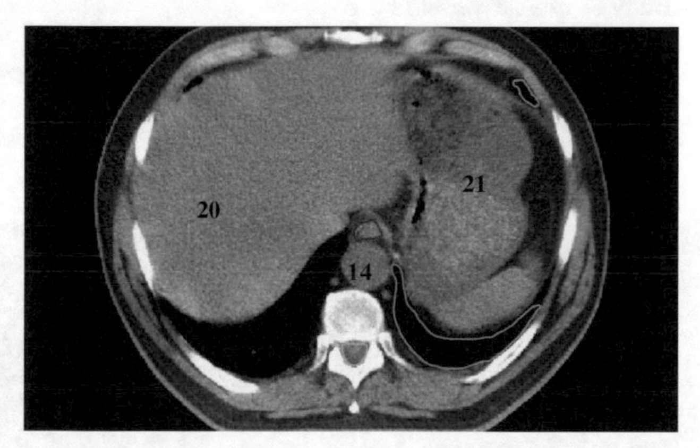

Fig. 5.58

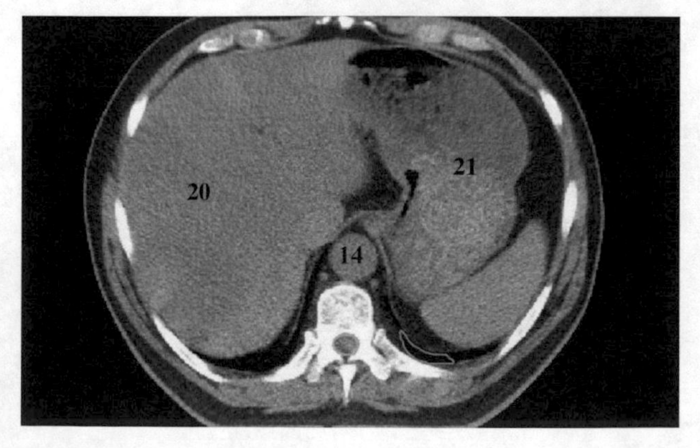

Fig. 5.59

Acknowledgement This chapter has been written with the contributions of Maria Taraborrelli, Rossella Patea, Lucia Anna Ursini, Monica Di Tommaso, and Marianna Tignani.

Anatomical Reference Points

1 – Heart
2 – Descending aorta
3 – Esophagus
4 – Inferior vena cava
5 – Spleen
6 – Diaphragm
7 – Body and tail of pancreas
8 – Hepatic hilum
9 – Gallbladder
10 – Pancreas (head)
11 – Body of lumbar vertebra (L3)

Anatomical Boundaries

ORGAN AT RISK	CRANIAL	CAUDAL	ANTERIOR
Liver	Right part of diaphragm	Variable (adipose tissue)	Ascending colon, lung, abdominal wall
Right kidney	Adrenal gland vena cava	Adipose tissue	Adipose tissue, ascending colon, duodenum
Left kidney	Adrenal gland, spleen	Adipose tissue	Body and tail of pancreas, adipose tissue
Stomach	Left part of diaphragm	Duodenum	Heart, adipose tissue, small intestine
Spinal cord	Occipital condyle	Inferior surface of L2	Vertebral canal
Small intestine	Angle of Treitz	Last ileal loop, ileocecal valve (right sacroiliac joint)	Variable
Ascending colon	Left hepatic lobe	Adipose tissue	Abdominal wall
Transverse colon	Stomach	Descending colon	Adipose tissue (omentum), abdominal wall
Descending colon	Body and tail of pancreas, spleen, kidney	Sigmoid colon, ileopsoas muscle	Transverse colon, adipose tissue
Sigmoid colon	Descending- sigmoid junction	Rectum - sigmoid junction	Small intestine, adipose tissue, bladder (male)

Chapter 5.3
Abdomen

Color Legend

- Spinal cord
- Liver
- Stomach
- Colon
- Left kidney
- Right kidney
- Duodenum
- Small intestine

POSTERIOR	MEDIAL	LATERAL	CT window
Lung, chest wall, adipose tissue, right kidney	Vena cava, Heart, esophagus, adipose tissue, gastric antrum, pancreas (head), ascending colon	Chest wall	Abdomen: C40, W400
Adipose tissue	Diaphragm, adipose tissue, iliopsoas muscle, hilum of kidney	Liver, adipose tissue, ascending colon	Abdomen: C40, W400
Adipose tissue	Diaphragm, iliopsoas muscle, adipose tissue, hilum of kidney	Spleen, adipose tissue, descending colon	Abdomen: C40, W400
Aorta, body and tail of pancreas, spleen (variable)	Abdominal aorta, diaphragm, body of pancreas, adipose tissue	Left hepatic lobe, adipose tissue, ascending colon	Abdomen: C40, W400
	Vertebral canal	vertebral canal	Bone: C450, W1600
Variable	Variable	Descending colon to the left, ascending colon to the right, adipose tissue	Abdomen: C40, W400
Kidney, adipose tissue, liver	Stomach, duodenum, adipose tissue, ileocecal valve	Liver, abdominal wall	Abdomen: C40, W400
Adipose tissue (transverse mesocolon), small intestine			Abdomen: C40, W400
Adipose tissue	Body and tail of pancreas, spleen, kidney, small intestine	Lateral abdomen wall	Abdomen: C40, W400
Sacrum	Uterum, left parametrium	Right parametrium, adipose tissue, iliac vessels (male)	Abdomen: C40, W400

G. Ausili Cefaro, D. Genovesi, C.A. Perez, *Delineating Organs at Risk in Radiation Therapy*,
DOI: 10.1007/978-88-470-5257-4_5-3, © Springer-Verlag Italia 2013

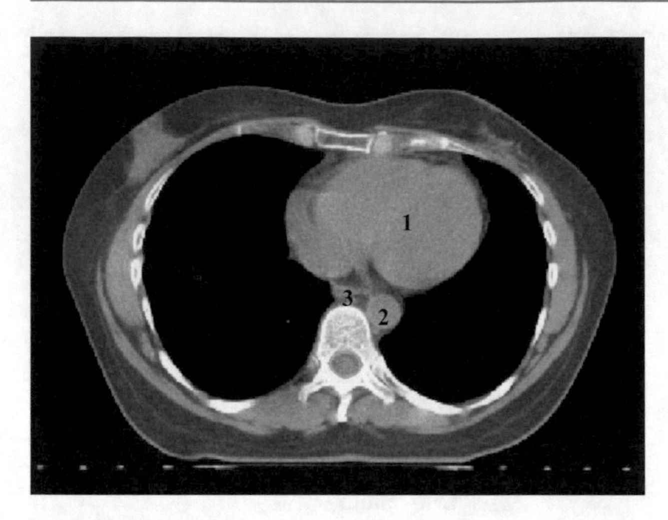

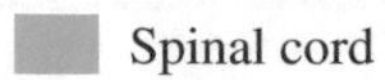
Spinal cord

Liver

1 – Heart

2 – Descending aorta

3 – Esophagus

Fig. 5.60

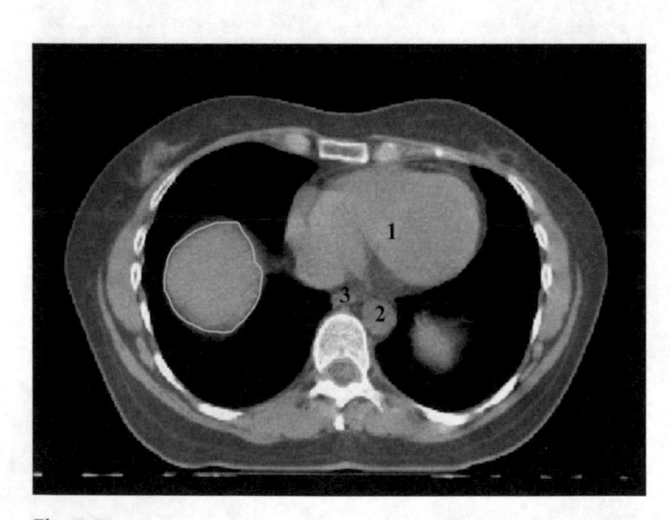

Fig. 5.61

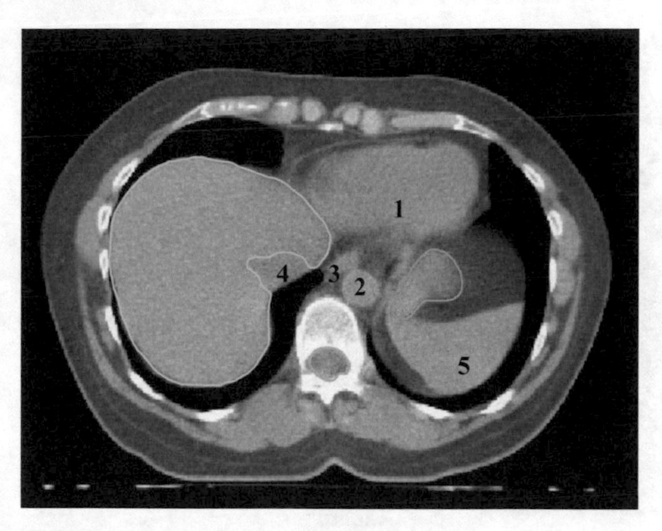

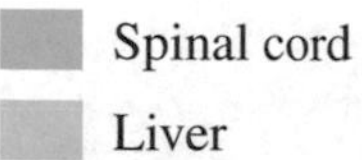

1 – Heart
2 – Descending aorta
3 – Esophagus
4 – Inferior vena cava
5 – Spleen

Fig. 5.62

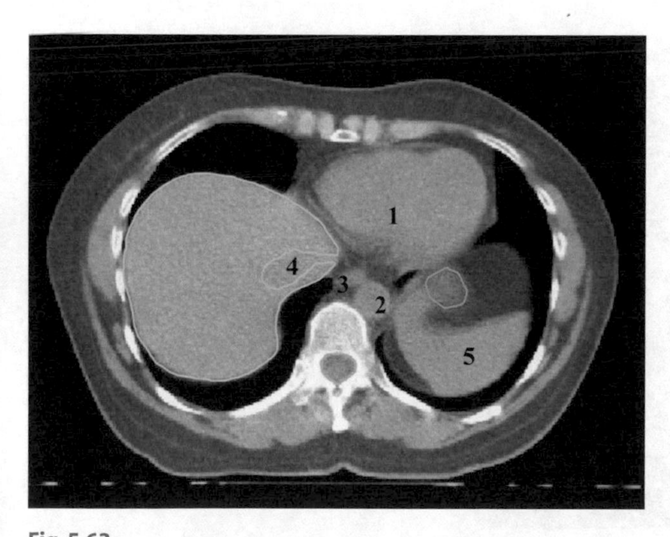

Fig. 5.63

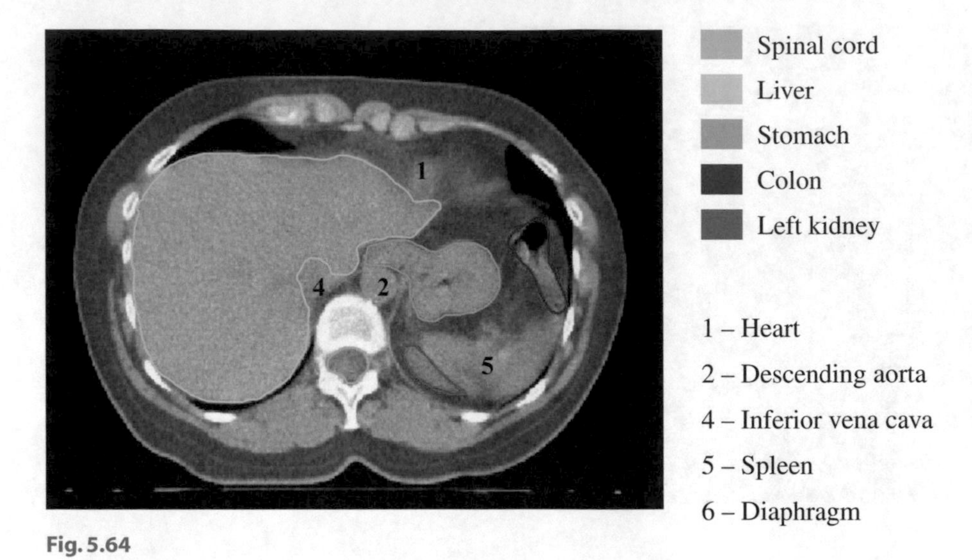

Fig. 5.64

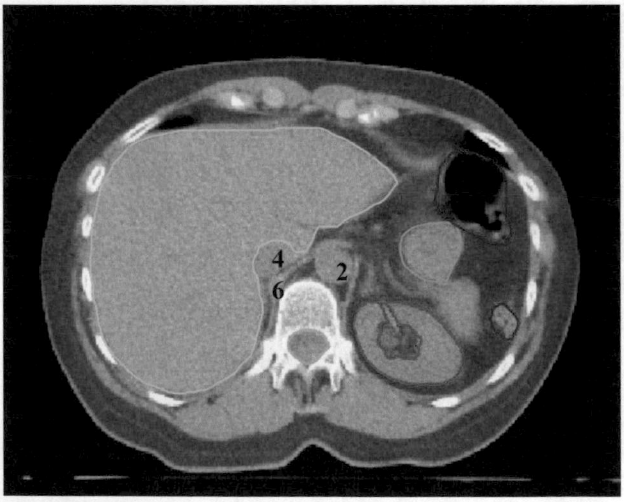

Fig. 5.65

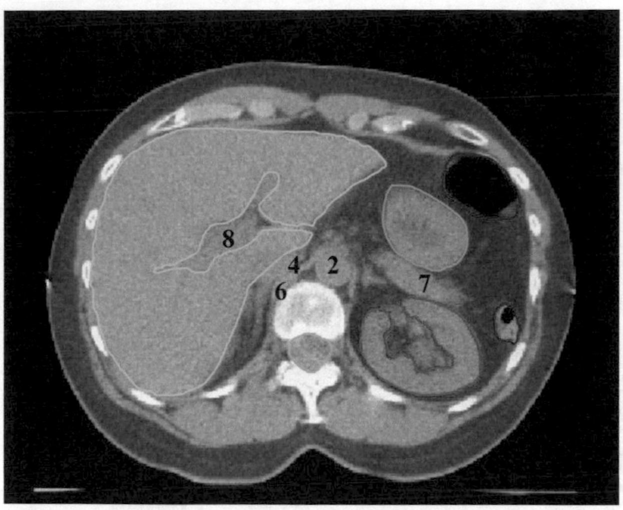

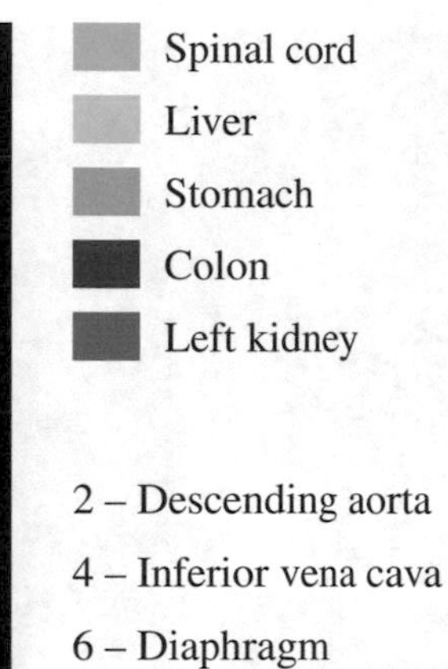

2 – Descending aorta

4 – Inferior vena cava

6 – Diaphragm

7 – Body and tail of pancreas

8 – Hepatic hilum

Fig. 5.66

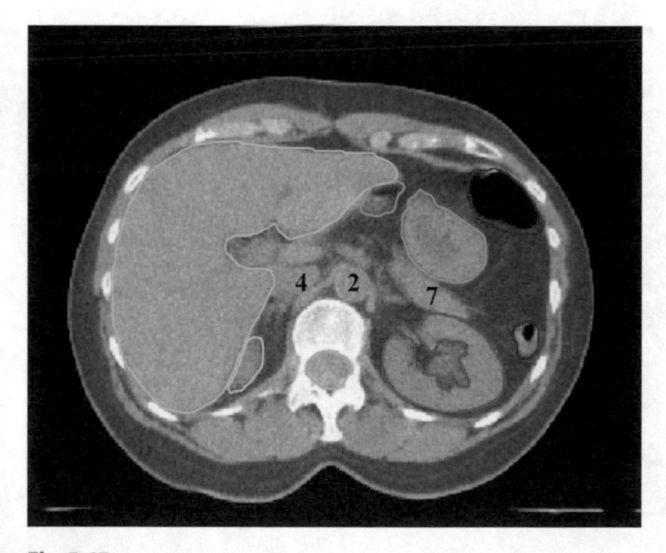

Fig. 5.67

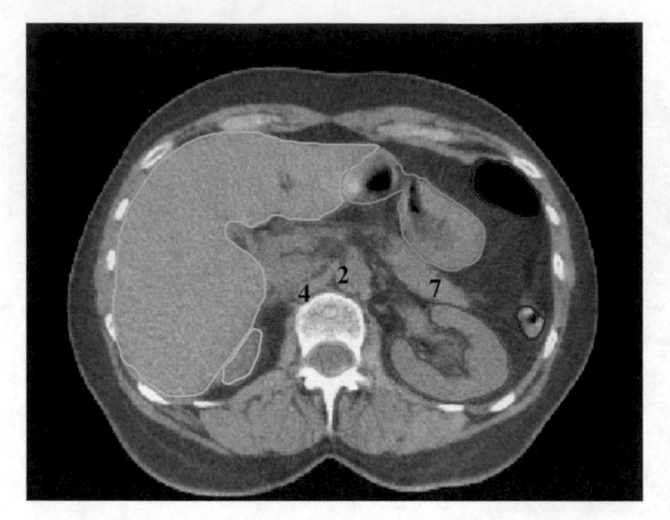

Fig. 5.68

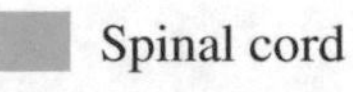

Spinal cord

Liver

Stomach

Colon

Left kidney

Right kidney

Duodenum

2 – Descending aorta

4 – Inferior vena cava

7 – Body and tail of pancreas

9 – Gallbladder

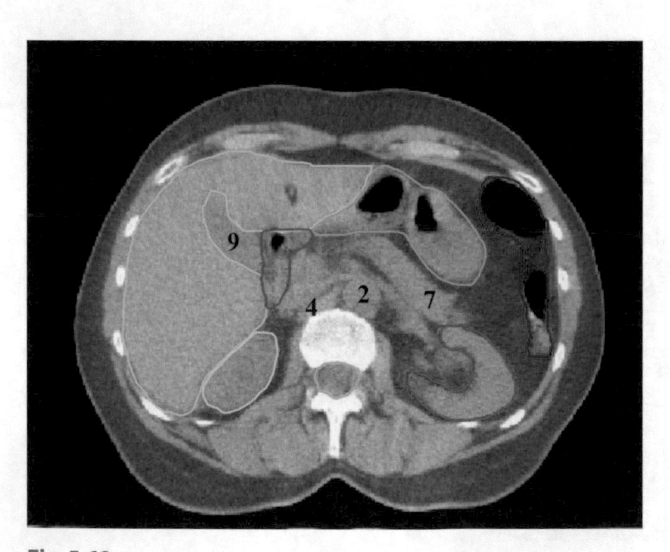

Fig. 5.69

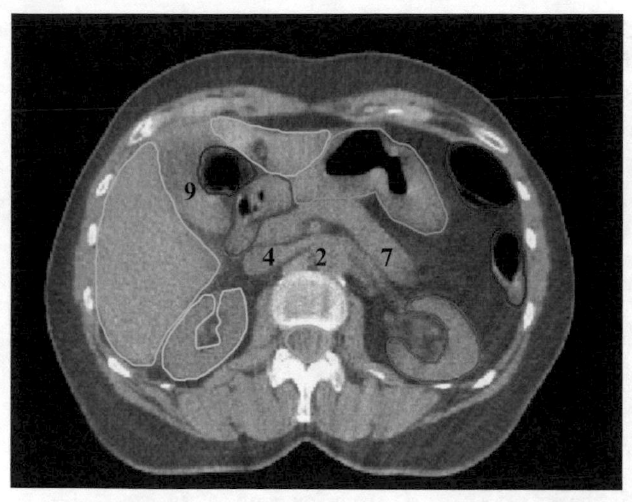

Fig. 5.70

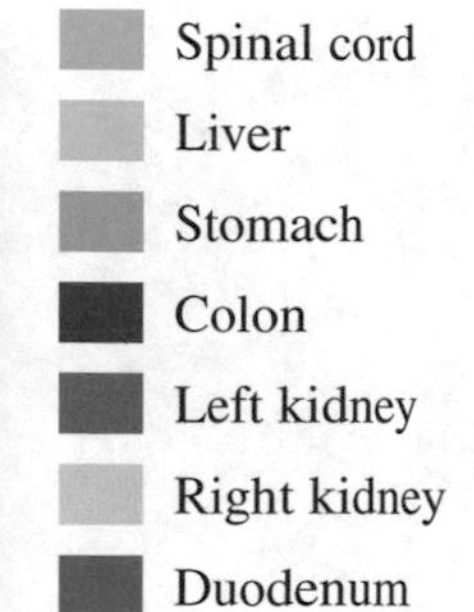

2 – Descending aorta

4 – Inferior vena cava

7 – Body and tail of pancreas

9 – Gallbladder

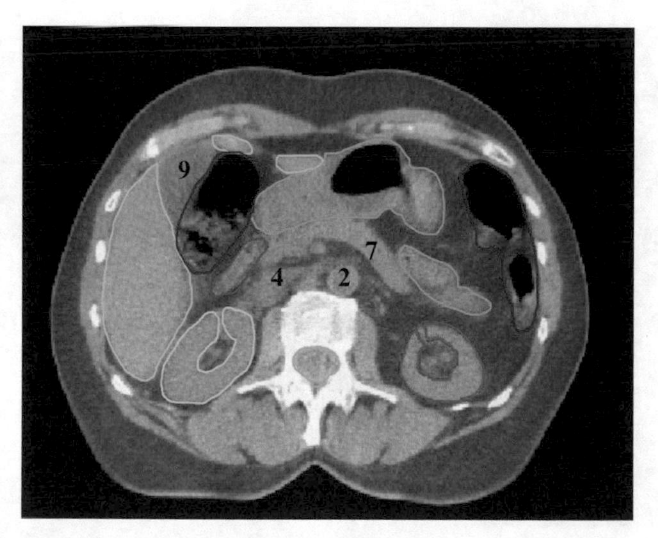

Fig. 5.71

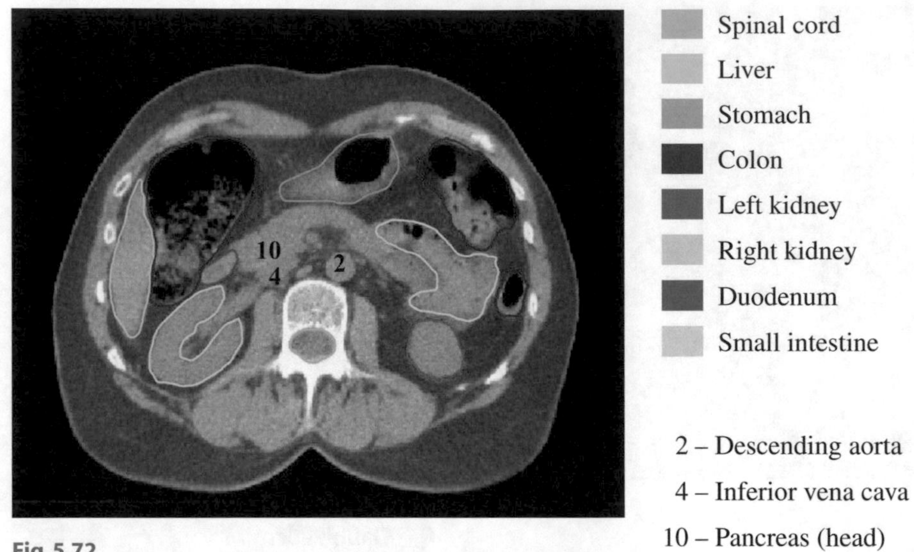

Fig. 5.72

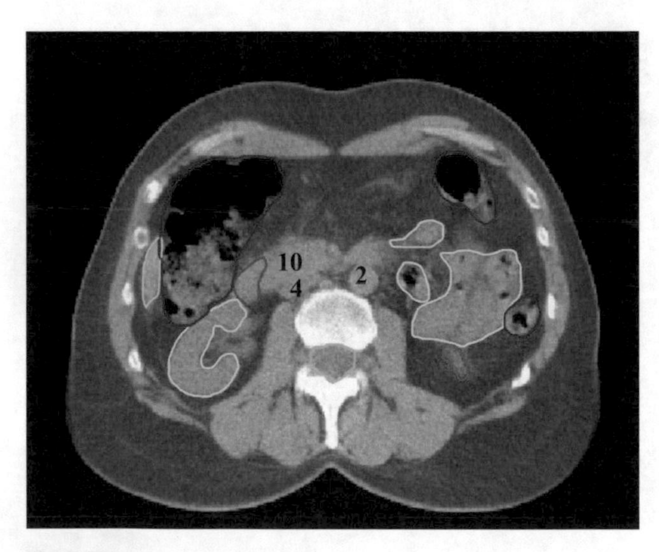

Fig. 5.73

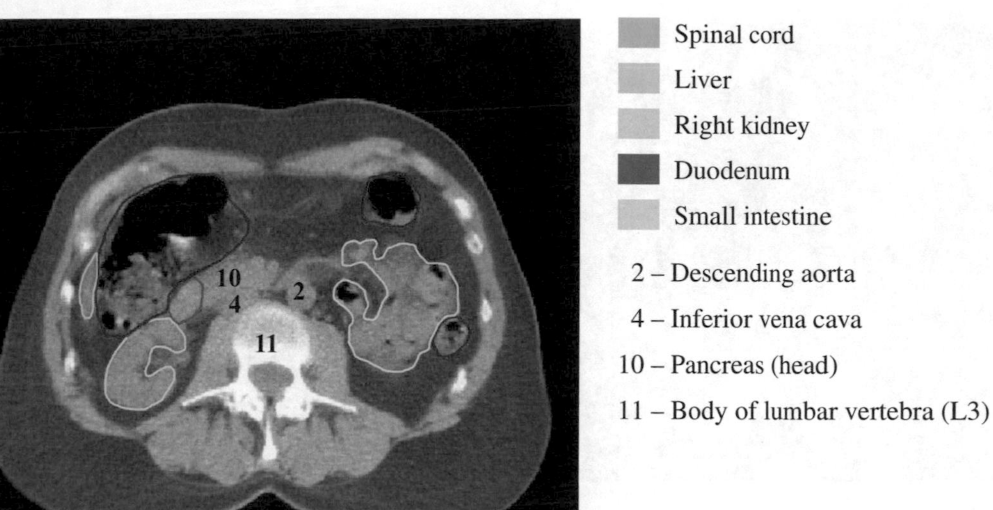

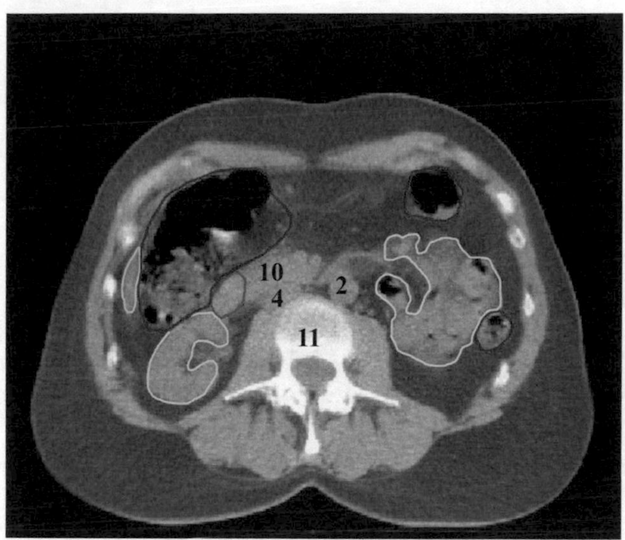

Fig. 5.74

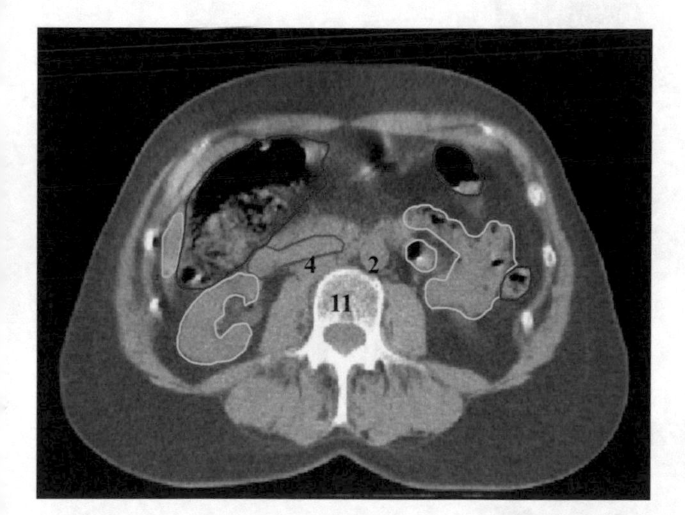

Fig. 5.75

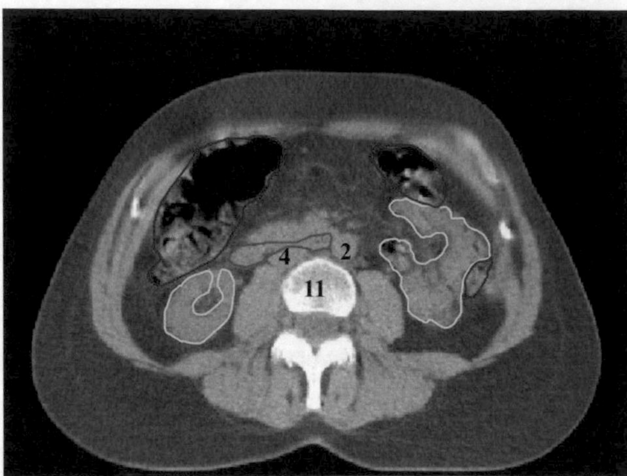
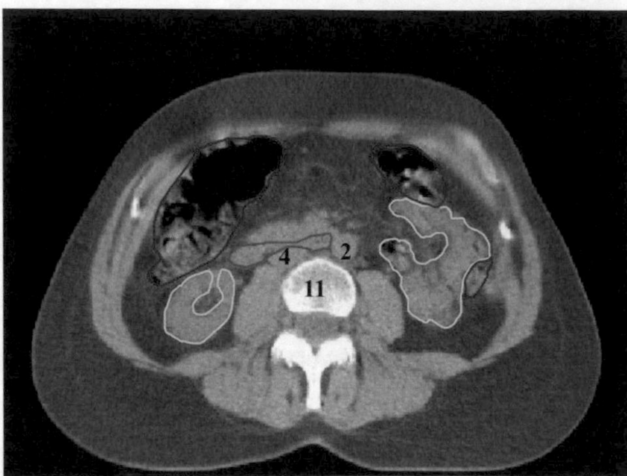

Fig. 5.76

Colon

Right kidney

Duodenum

Small intestine

2 – Descending aorta

4 – Inferior vena cava

11 – Body of lumbar vertebra (L3)

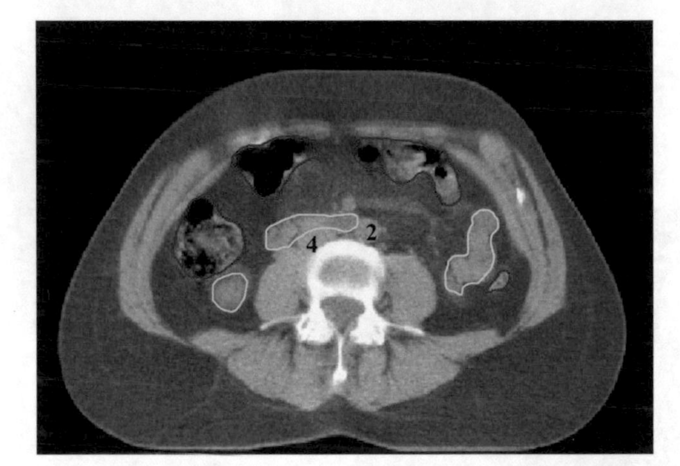

Fig. 5.77

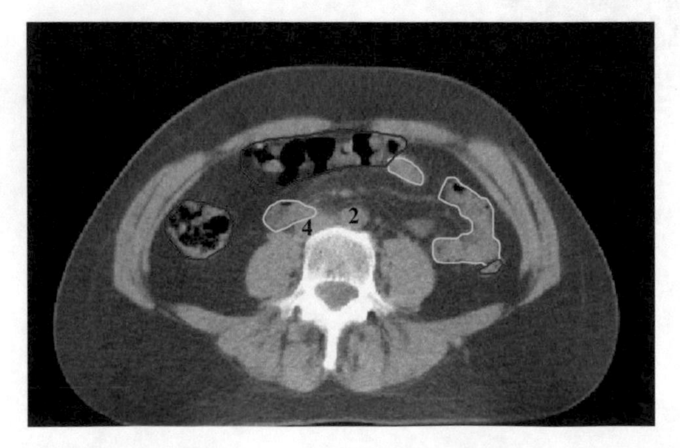

Fig. 5.78

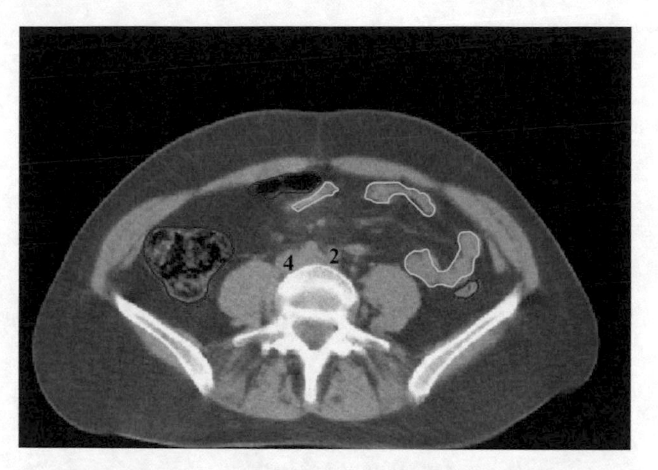

Fig. 5.79

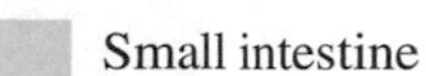

Colon

Small intestine

2 – Descending aorta

4 – Inferior vena cava

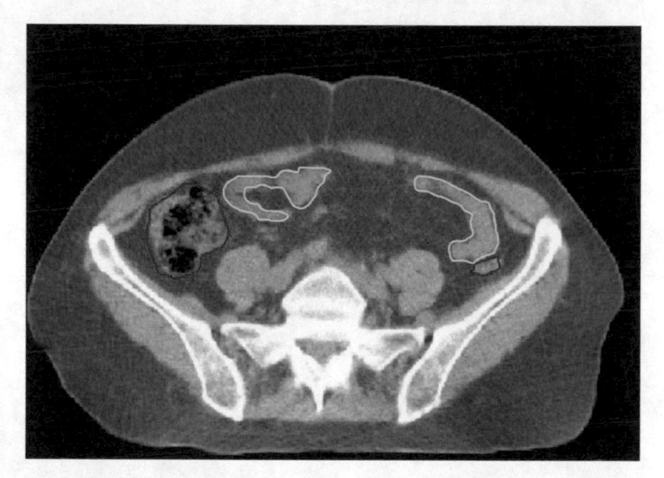

Fig. 5.80

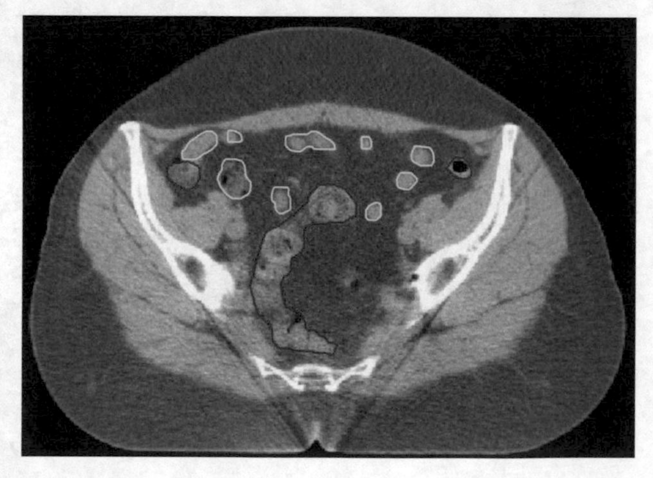

Fig. 5.81

Acknowledgement This chapter has been written with the contributions of Antonietta Augurio, Raffaella Basilico, Marianna Trignani, and Monica Di Tommaso.

Anatomical Reference Points

1 – Sigmoid colon/bladder
2 – Rectus abdominis muscles
3 – External iliac vessels
4 – Sacrum
5 – Seminal vesicles
6 – Prostate
7 – Pubic symphysis
8 – Levator ani muscle
9 – Prostatic apex
10 – External anal sphincter
11 – Penis origin
12 – Ischiocavernosus muscle
13 – Smaller trochanter

Anatomical Boundaries

ORGAN AT RISK	CRANIAL	CAUDAL	ANTERIOR
Penile bulb	1 cm below the inferior margin of the symphysis	Inferior margin of smaller trochanter, tangent plane to the lower profile of pubic bone (inferior ramus), corpus cavernosum	Corpus cavernosum
Bladder	Small intestine (variable level depending on the filling)	Inferior margin of the symphysis	Small intestine, rectus abdominis muscles, symphysis (superior margin)
Rectum	Recto–sigmoid junction	Inferior margin of the external anal sphincter	Prostate, penile bulb
Urogenital diaphragm	Inferior margin of the symphysis, ischiopubic bone, prostatic apex	1–1.5 cm from the edge of the cranial	Penis origin
Head of femur	Lower edge of the acetabulum	Lower limit of the acetabulum	Acetabulum

Chapter 5.4
Male Pelvis

Color Legend

- Bladder
- Rectum
- Head of right femur
- Head of left femur
- Urogenital diaphragm
- Penile bulb

POSTERIOR	MEDIAL	LATERAL	CT window
Anal canal		Ischiocavernosus muscles	Pelvis C250 W1,000
Sigmoid colon, seminal vesicles, prostate		External iliac vessels	Pelvis C250 W1,000
Sacrum, levator ani muscle		Levator ani muscle, external anal sphincter	Pelvis C250 W1,000
Levator ani muscle, external anal sphincter, anal canal		Ischiocavernosus muscle	Pelvis C250 W1,000
Acetabulum	Acetabulum	Femoral neck	Bone C450 W1,600

G. Ausili Cefaro, D. Genovesi, C.A. Perez, *Delineating Organs at Risk in Radiation Therapy*,
DOI: 10.1007/978-88-470-5257-4_5-4, © Springer-Verlag Italia 2013

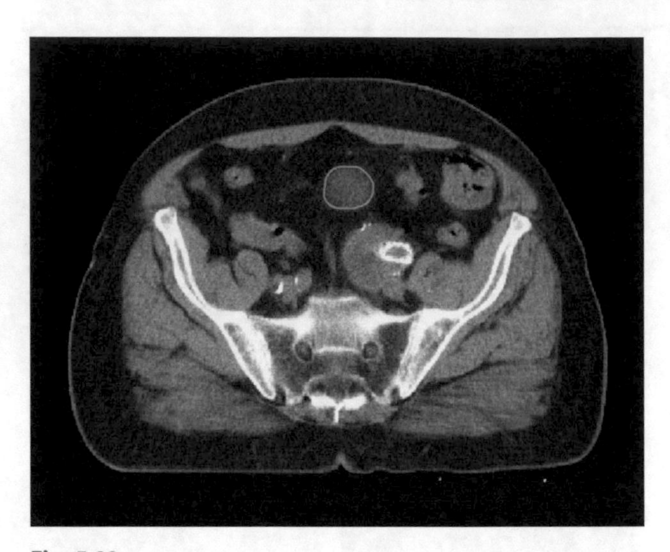

Fig. 5.82

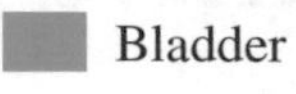

Bladder

1 – Sigmoid colon/bladder

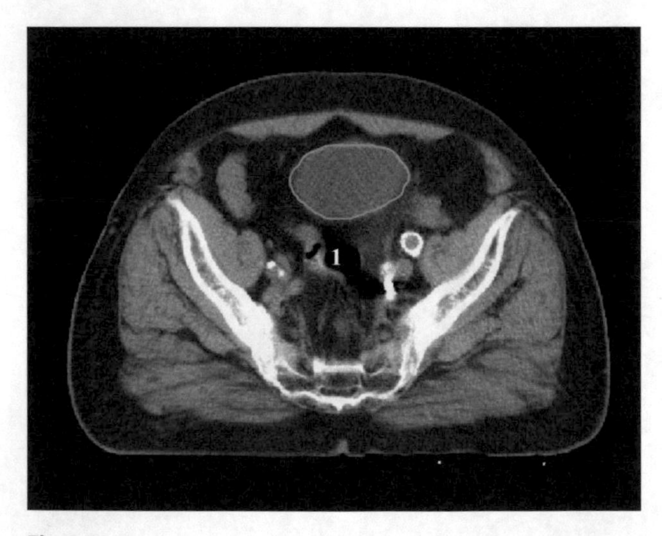

Fig. 5.83

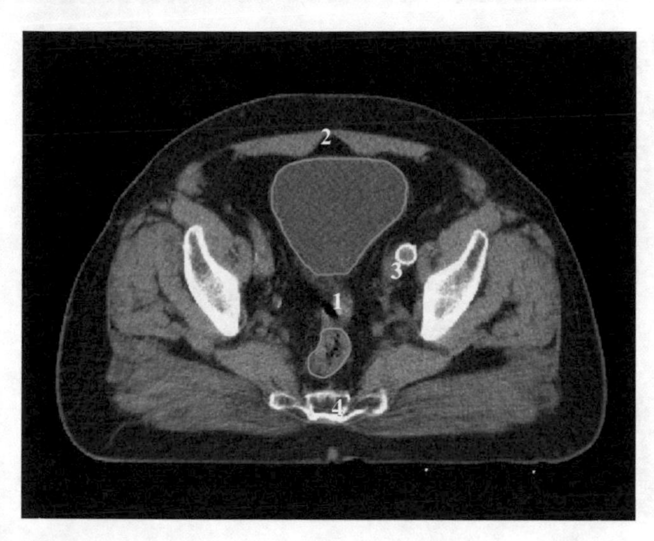

1 – Sigmoid colon/bladder

2 – Rectus abdominis muscle

3 – External iliac vessels

4 – Sacrum

Fig. 5.84

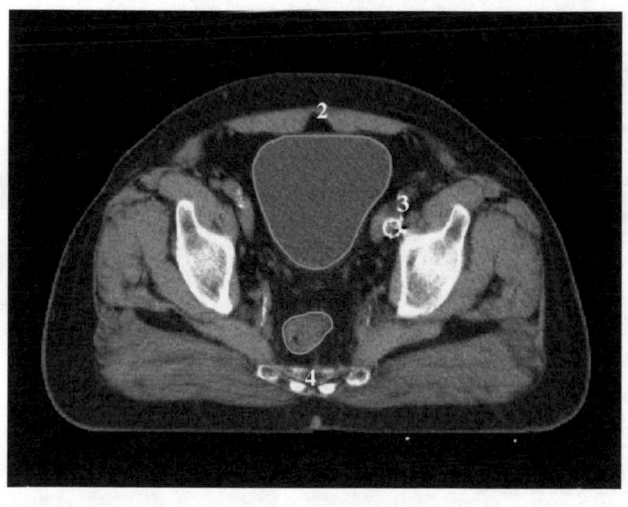

Fig. 5.85

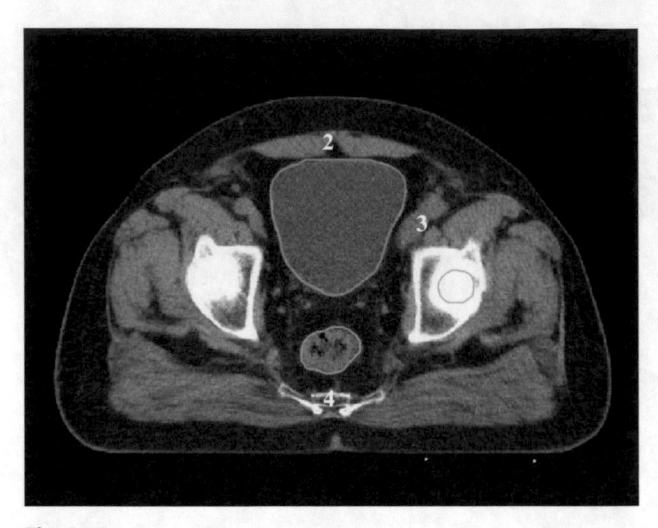

Bladder
Rectum
Head of right femur
Head of left femur

2 – Rectus abdominis muscle
3 – External iliac vessels
5 – Seminal vesicles
4 – Sacrum

Fig. 5.86

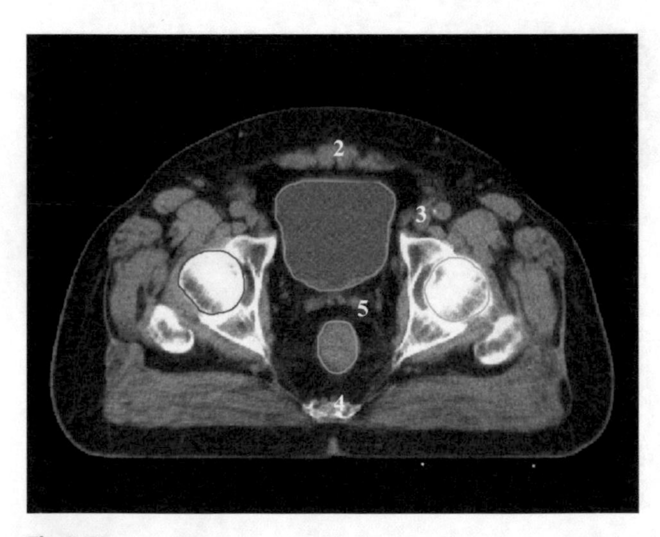

Fig. 5.87

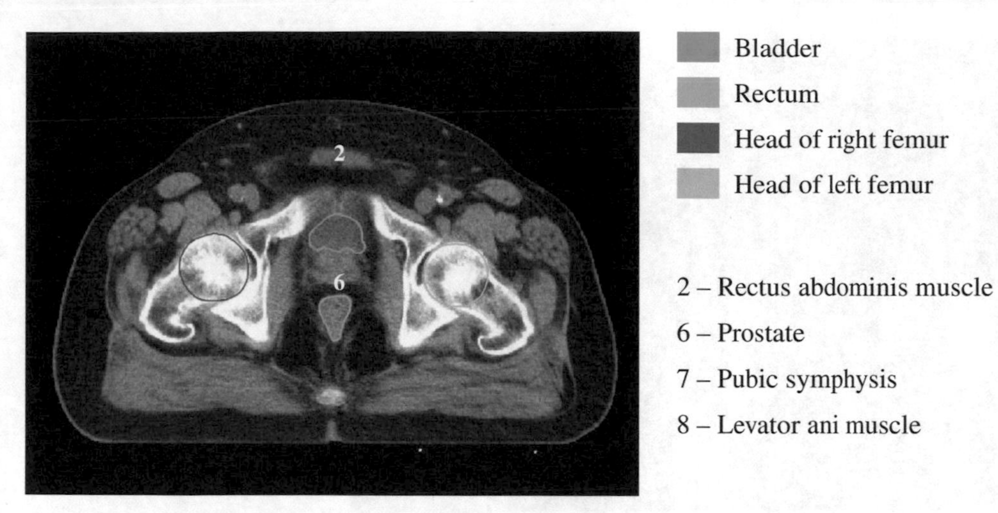

Fig. 5.88

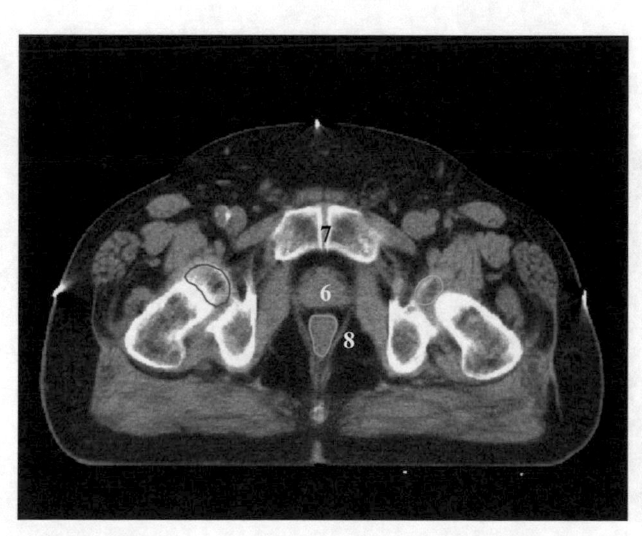

Fig. 5.89

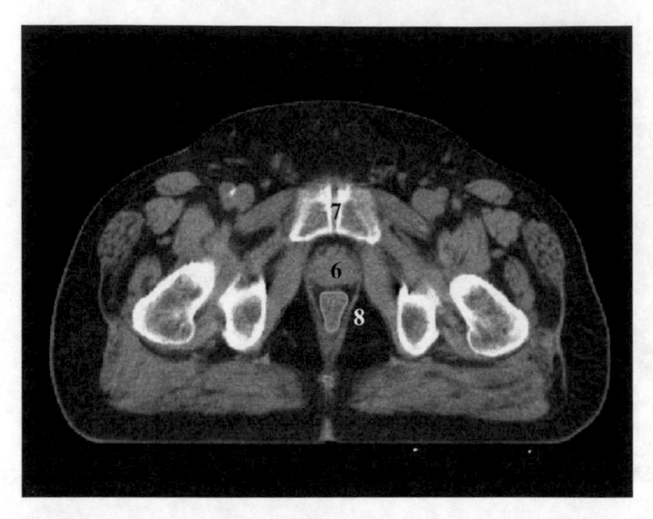

 Rectum

6 – Prostate

7 – Pubic symphysis

8 – Levator ani muscle

9 – Prostatic apex

Fig. 5.90

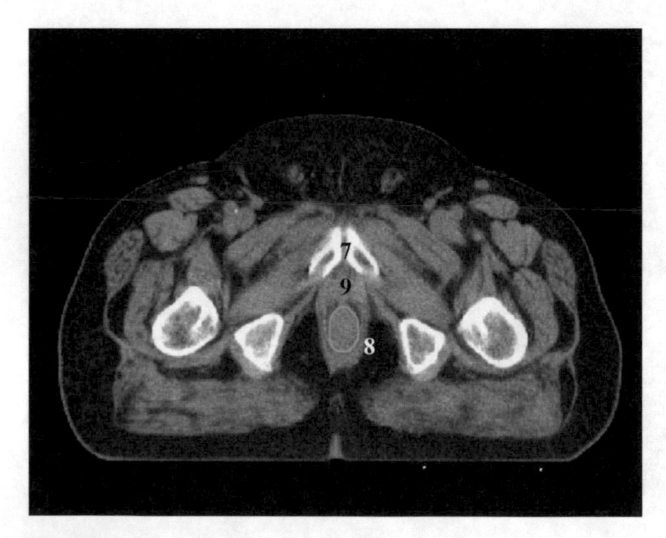

Fig. 5.91

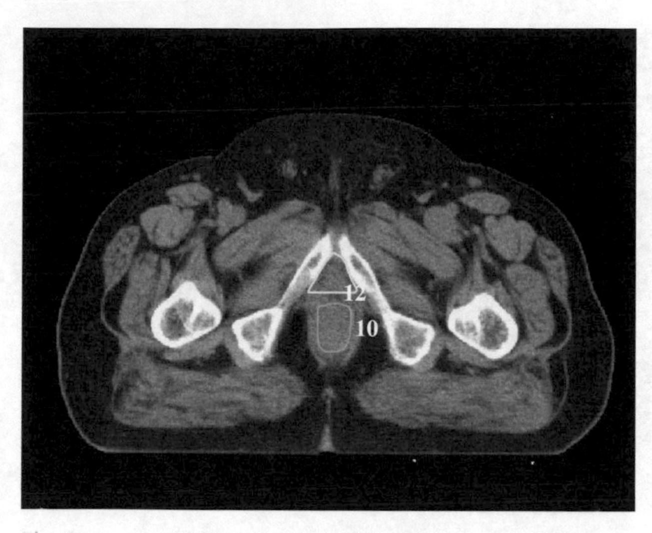

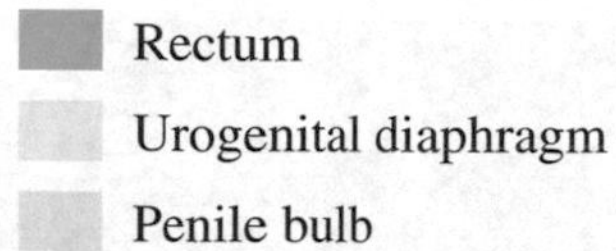

10 – External anal sphincter

11 – Penis origin

12 – Ischiocavernosus muscle

Fig. 5.92

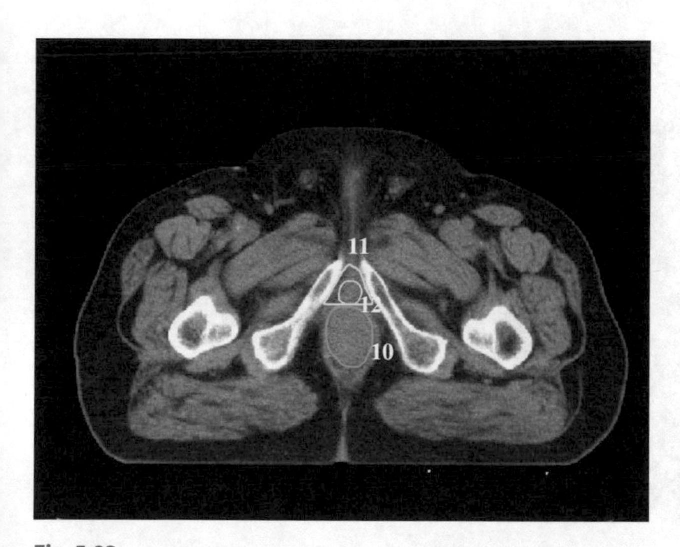

Fig. 5.93

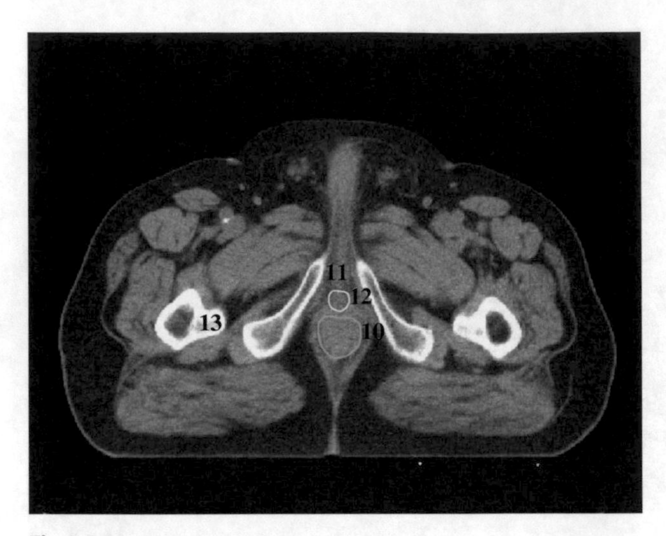

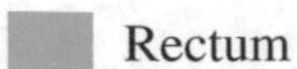
Rectum

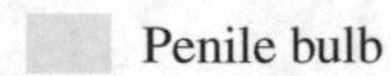
Penile bulb

10 – External anal sphincter

11 – Penis origin

12 – Ischiocavernosus muscle

13 – Smaller trochanter

Fig. 5.94

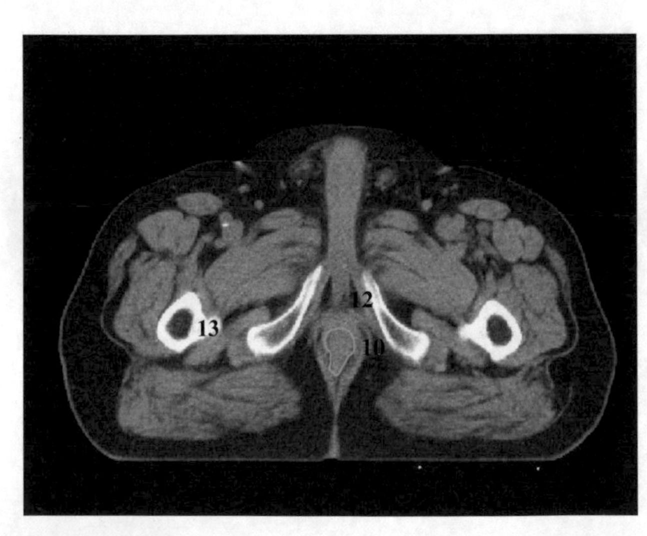

Fig. 5.95

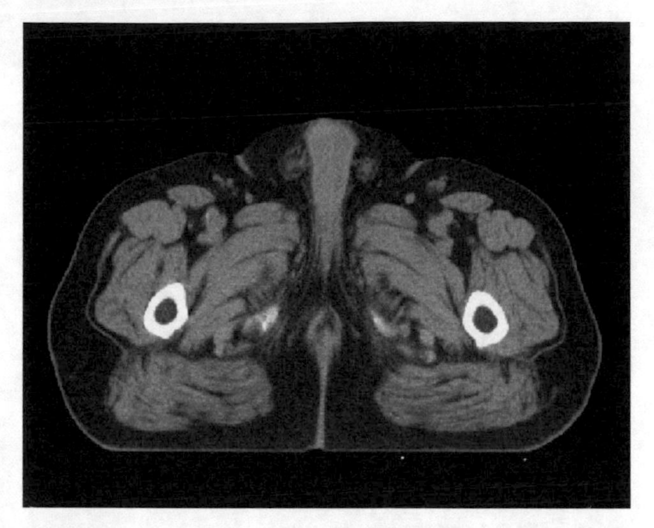

Fig. 5.96

Acknowledgement This chapter has been written with the contributions of Antonella Filippone, Marianna Trignani, and Monica Di Tommaso.

Anatomical Reference Points

1 – Sigmoid colon
2 – External iliac vessels
3 – Iliopsoas muscle
4 – Internal iliac vessels
5 – Sacrum
6 – Rectum abdominis muscles
7 – Piriformis muscle
8 – Uterus
9 – Vagina
10 – Levator ani muscle
11 – Pubic symphysis
12 – Ischiocavernosus muscle
13 – External levator ani muscle

Anatomical Boundaries

ORGAN AT RISK	CRANIAL	CAUDAL	ANTERIOR
Bladder	Small intestine (variable level depending on the filling)	Inferior margin of the symphysis	Small intestine, rectus abdominis muscle, symphysis (superior margin)
Rectum	Recto–sigmoid junction	Inferior margin of the external anal sphincter	Uterus, vagina
Urogenital diaphragm	Inferior margin of the symphysis, ischiopubic bone, prostatic apex	1–1.5 cm from the edge of the cranium	
Head of femur	Lower edge of the acetabulum	Lower limit of the acetabulum	Acetabulum
Ovary		Lower edge of the acetabulum	External iliac vessels

Chapter 5.5
Female Pelvis

Color Legend

- Left ovary
- Right ovary
- Rectum
- Bladder
- Head of right femur
- Head of left femur
- Urogenital diaphragm

POSTERIOR	MEDIAL	LATERAL	CT window
Sigmoid colon, uterus, vagina		External iliac vessels	Pelvis C250 W1,000
Sacrum, levator ani muscle		Levator ani muscle, external anal sphincter	Pelvis C250 W1,000
Levator ani muscle, external anal sphincter, anal canal		Ischiocavernosus muscles	Pelvis C250 W1,000
Acetabulum	Acetabulum	Femoral neck	Bone C450 W1,600
Internal iliac vessels, piriformis muscle	Bladder, uterus, sigmoid colon	Iliacus muscle	Pelvis C250 W1,000

G. Ausili Cefaro, D. Genovesi, C.A. Perez, *Delineating Organs at Risk in Radiation Therapy*,
DOI: 10.1007/978-88-470-5257-4_5-5, © Springer-Verlag Italia 2013

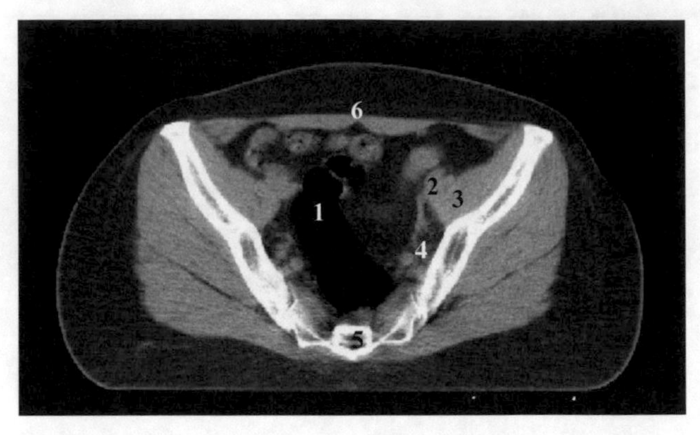

Fig. 5.97

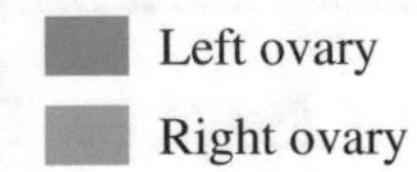

1 – Sigmoid colon

2 – External iliac vessels

3 – Iliopsoas muscle

4 – Internal iliac vessels

5 – Sacrum

6 – Rectum abdominis muscle

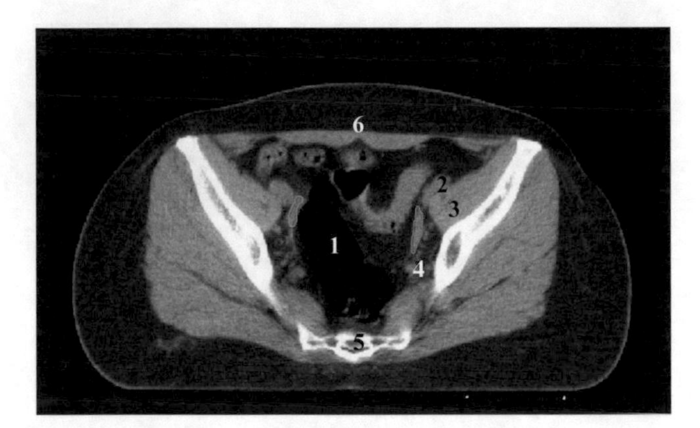

Fig. 5.98

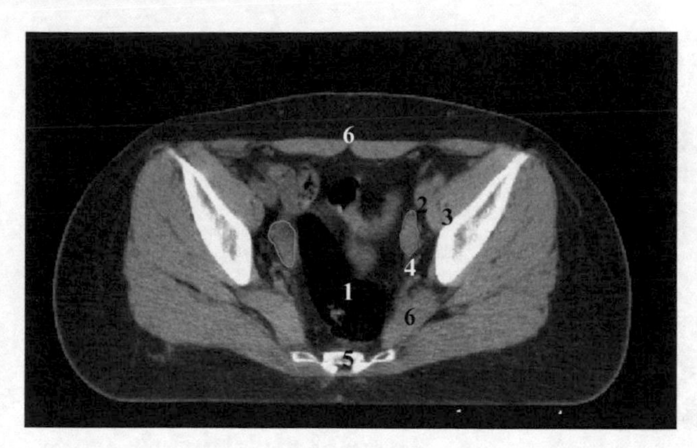

Fig. 5.99

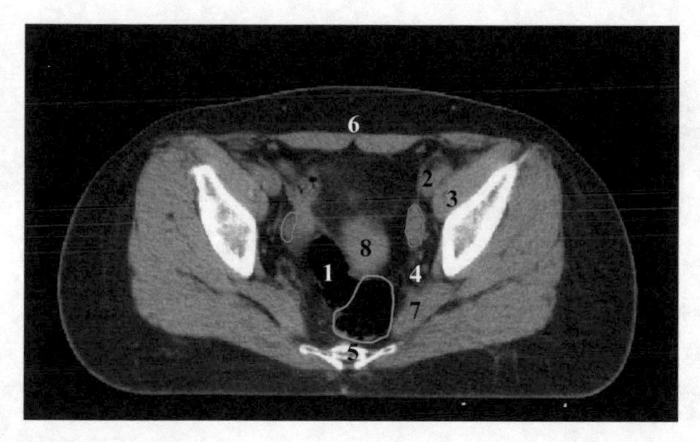

Fig. 5.100

Left ovary

Right ovary

Rectum

1 – Sigmoid colon

2 – External iliac vessels

3 – Iliopsoas muscle

4 – Internal iliac vessels

5 – Sacrum

6 – Rectum abdominis muscle

7 – Piriformis muscle

8 – Uterus

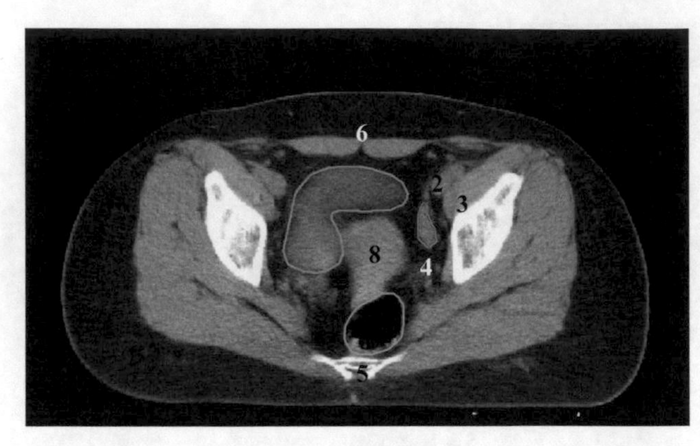

Fig. 5.101

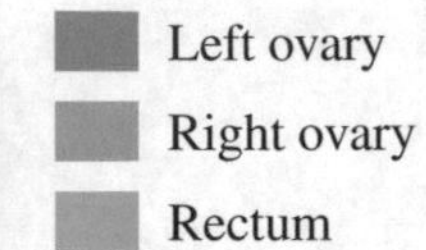

2 – External iliac vessels

3 – Iliopsoas muscle

4 – Internal iliac vessels

5 – Sacrum

6 – Rectum abdominis muscle

8 – Uterus

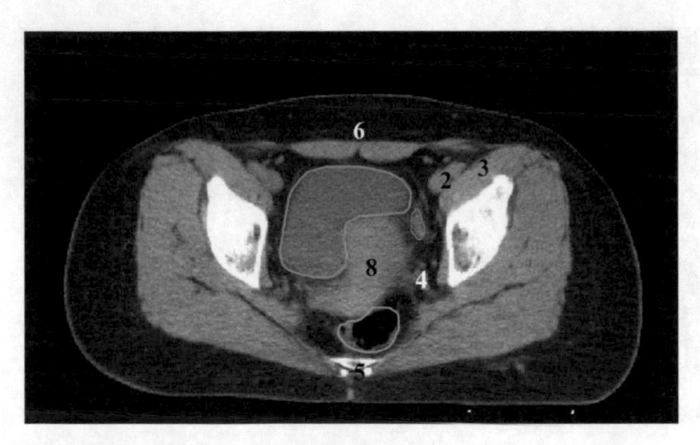

Fig. 5.102

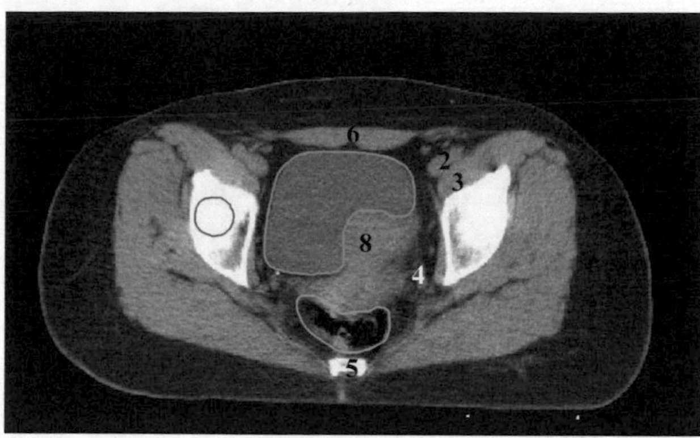

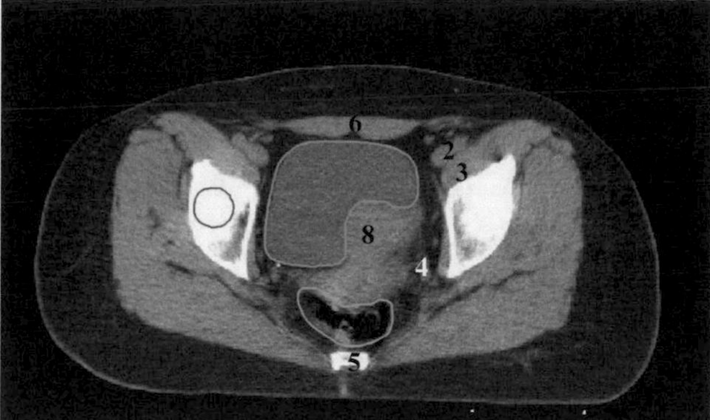

Fig. 5.103

Bladder

Rectum

Head of right femur

Head of left femur

2 – External iliac vessels

3 – Iliopsoas muscle

4 – Internal iliac vessels

5 – Sacrum

6 – Rectum abdominis muscle

8 – Uterus

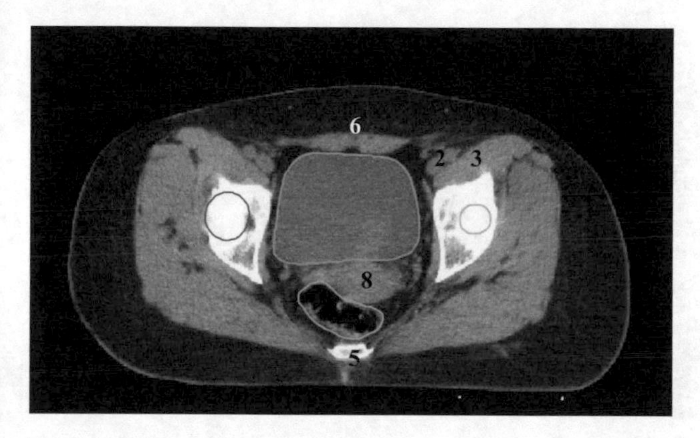

Fig. 5.104

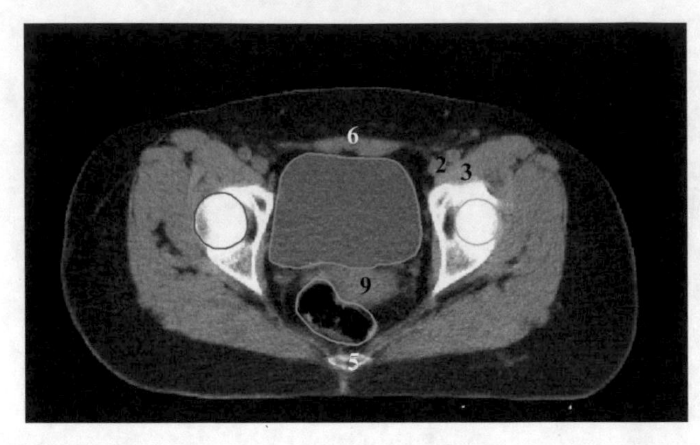

Fig. 5.105

Bladder

Rectum

Head of right femur

Head of left femur

2 – External iliac vessels

3 – Iliopsoas muscle

5 – Sacrum

6 – Rectum abdominis muscle

9 – Vagina

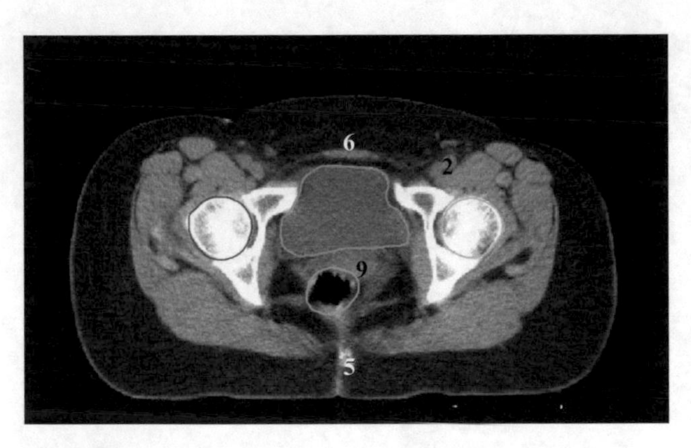

Fig. 5.106

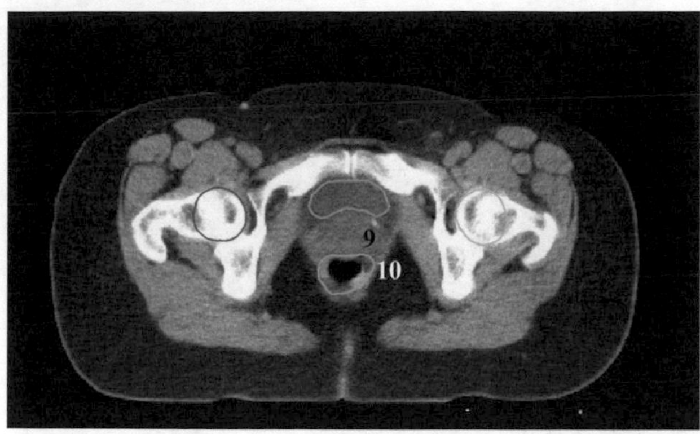

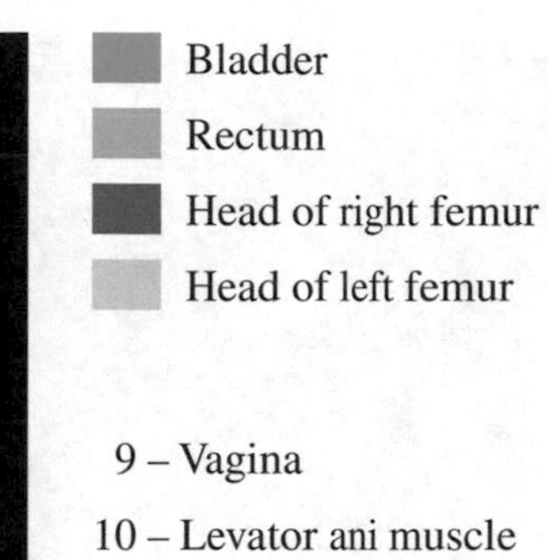

9 – Vagina

10 – Levator ani muscle

Fig. 5.107

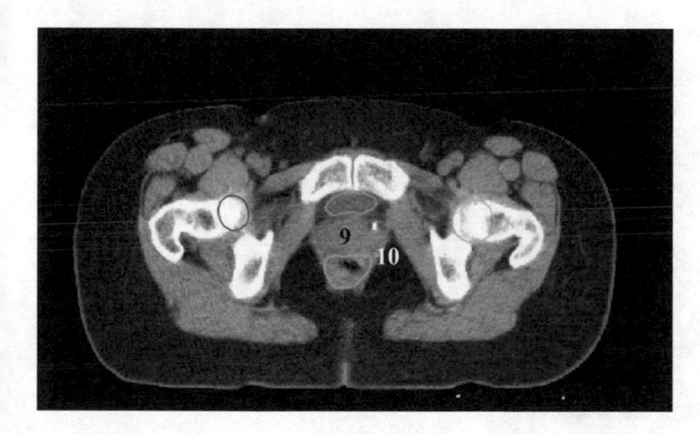

Fig. 5.108

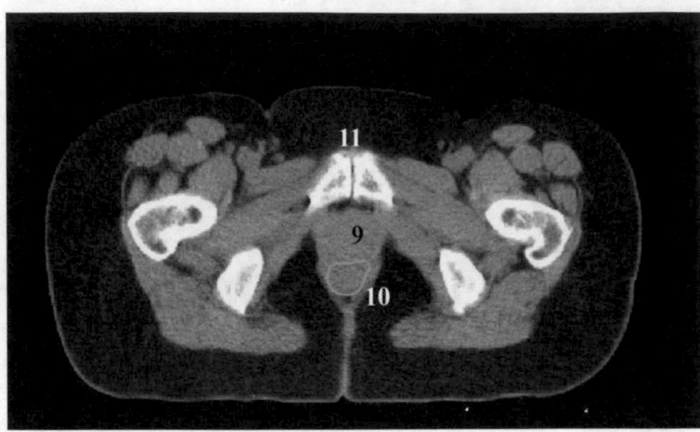

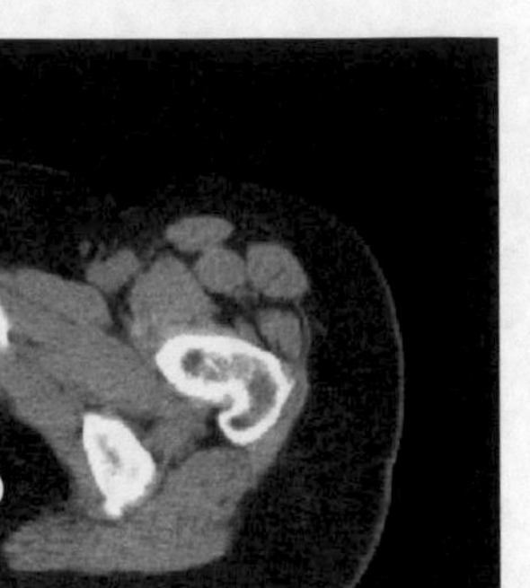

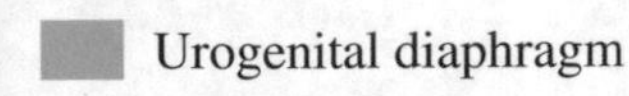

Rectum

Urogenital diaphragm

9 – Vagina

10 – Levator ani muscle

11 – Pubic symphysis

12 – Ischiocavernosus muscle

Fig. 5.109

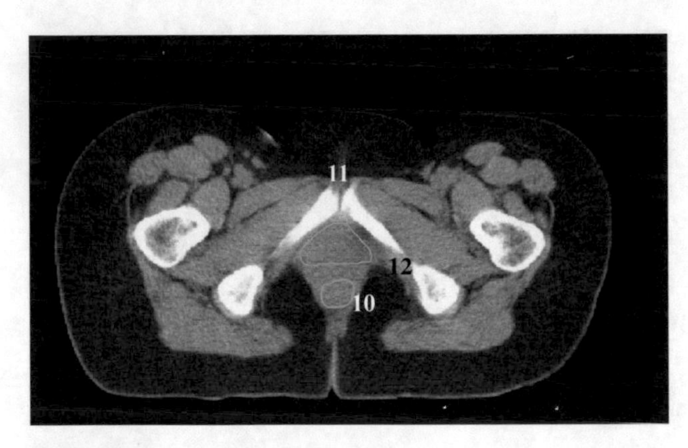

Fig. 5.110

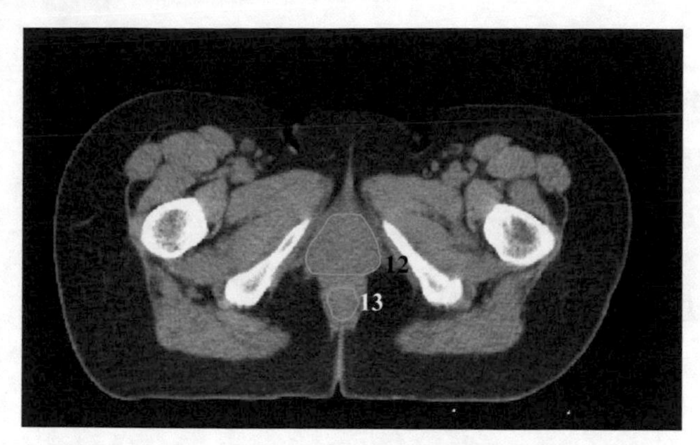

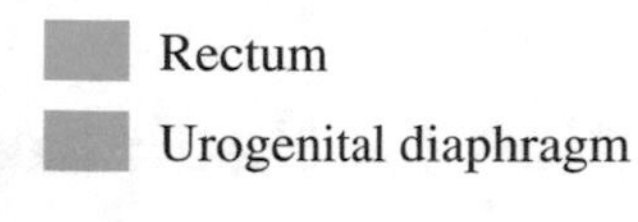

12 – Ischiocavernosus muscle

13 – External levator ani muscle

Fig. 5.111

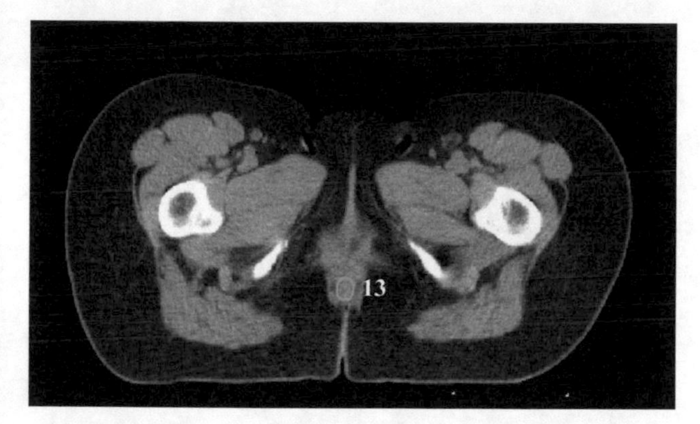

Fig. 5.112

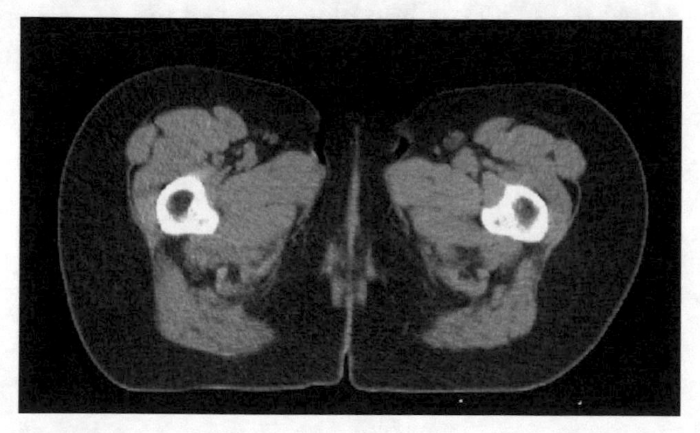

Fig. 5.113

Acknowledgement This chapter has been written with the contributions of Antonella Filippone, Marianna Trignani, and Monica Di Tommaso.

Correction to: Radiation Dose Constraints for Organs at Risk: Modeling and Importance of Organ Delineation in Radiation Therapy

G.A. Cefaro, D. Genovesi and C.A. Perez

Correction to: G.A. Cefaro, D. Genoveis and C.A. Perez, *Radiation Dose Constraints for Organs at Risk: Modeling and Importance of Organ Delineation in Radiation Therapy*,
DOI:10.1007/978-88-470-5257-4_6

The word "normal standard dose" was incorrectly captured in the second paragraph of 3.2 Synopsis of Historical Perspective, p. 50 of this book. The word should read as "nominal standard dose". This has been corrected.

The updated online version of the original chapter can be found at https://doi.org/10.1007/978-88-470-5257-4_6

G.A. Cefaro et al., *Delineating Organs at Risk in Radiation Therapy*,
DOI 10.1007/978-88-470-5257-4_13

Appendix
Digitally Reconstructed Radiographs

Some examples of digitally reconstructed radiographic images (DRRs) for different organs at risk and districts are reported below (Figs. 6.1-6.4).

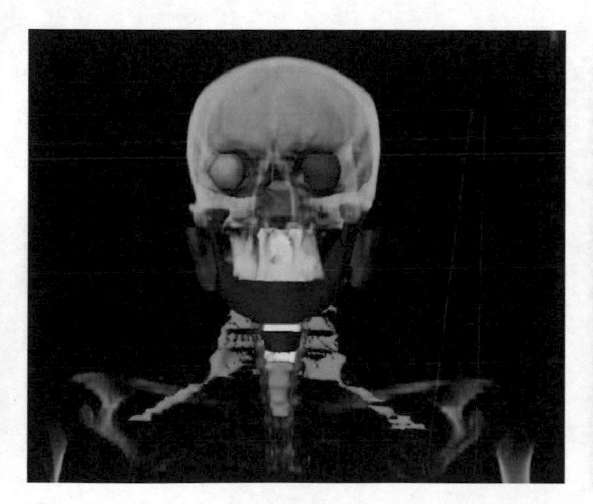

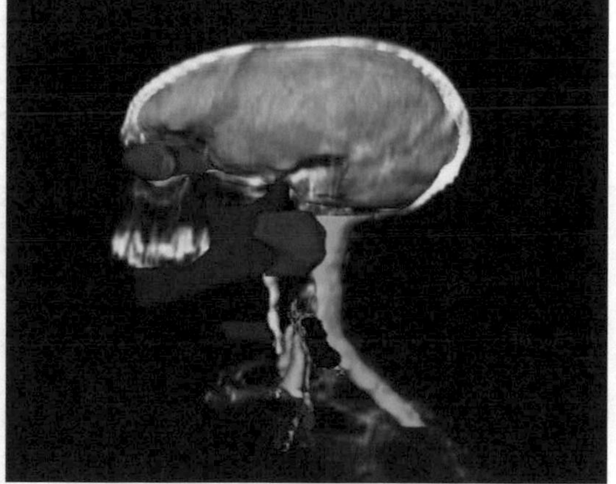

Fig. 6.1a,b Brain, head and neck, frontal and lateral view

G. Ausili Cefaro, D. Genovesi, C.A. Perez, *Delineating Organs at Risk in Radiation Therapy*,
DOI: 10.1007/978-88-470-5257-4_6,

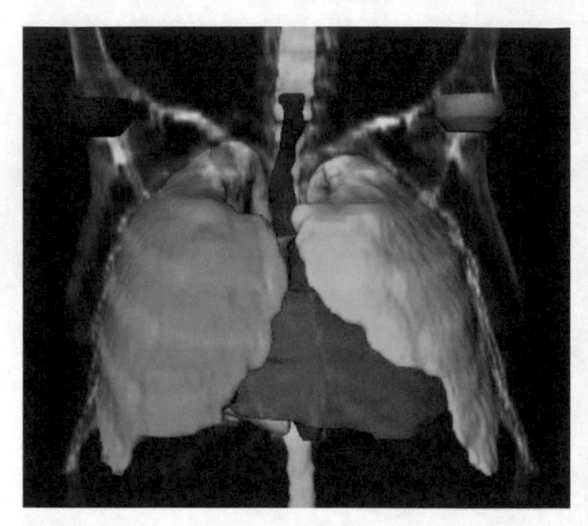

Fig. 6.2 Mediastinum, frontal view

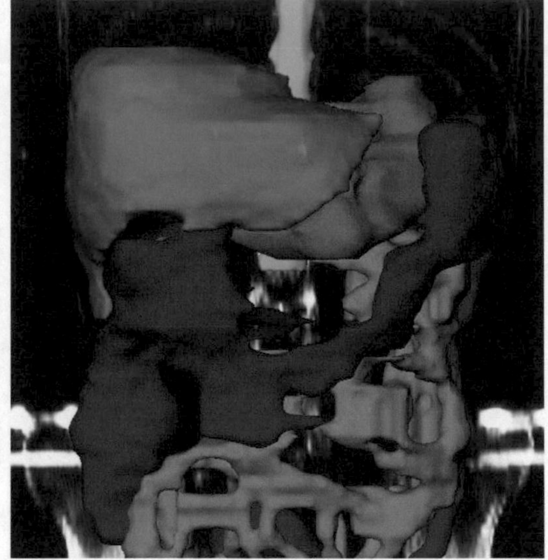

Fig. 6.3 Abdomen, frontal view

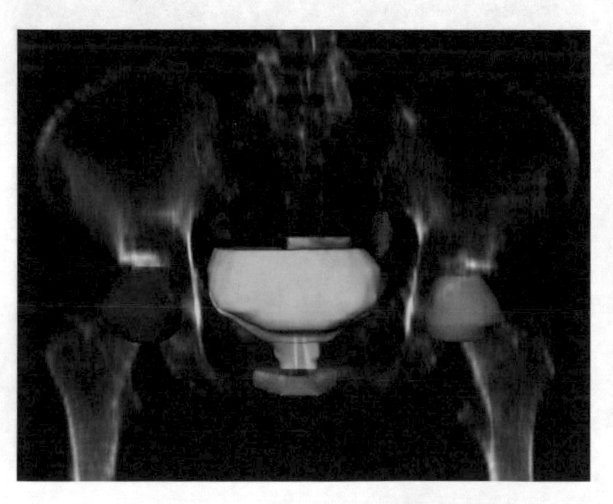

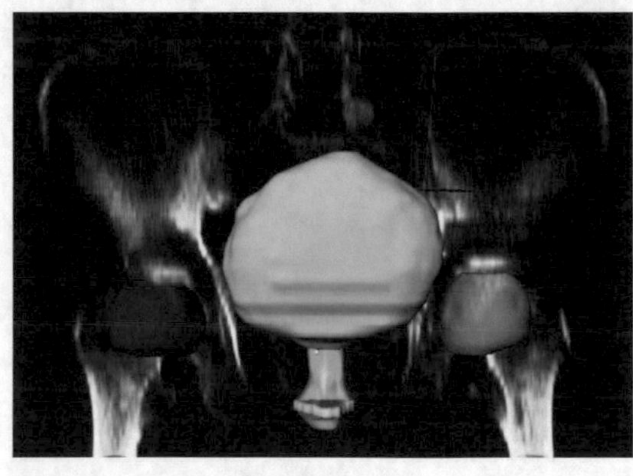

Fig. 6.4a,b Female and male pelvis, frontal view

Acknowledgement This chapter has been written with the contribution of Monica Di Tommaso.

Printed in the United States
by Baker & Taylor Publisher Services